Electronic Measurements and Testing

Tips and Techniques for Technicians and Engineers

Eugene R. Bartlett, P.E.

D1127911

McGraw-Hill, Inc.

New York   St. Louis   San Francisco   Auckland   Bogotá
Caracas   Lisbon   London   Madrid   Mexico   Milan
Montreal   New Delhi   Paris   San Juan   São Paulo
Singapore   Sydney   Tokyo   Toronto

Library of Congress Cataloging-in-Publication Data

Bartlett, Eugene R.
    Electronic measurements and testing tips and techniques for
technicians and engineers / Eugene R. Bartlett
        p.   cm.
    Includes index.
    ISBN 0-07-003961-5 (hard)—ISBN 0-07-003692-3 (soft)
    1. Electric testing 2. Electric measurements. I. Title.
TK401.B37   1992
621.381'548—dc20   91-40005
                                           CIP

1 2 3 4 5 6 7 8 9 0 DOC/DOC 9 8 7 6 5 4 3 2 1

ISBN   0-07-003961-5
ISBN   0-07-003692-3   {PBK}

*The sponsoring editor for this book was Daniel A. Gonneau, the
editing supervisor was Kimberly A. Goff, and the production
supervisor was Suzanne W. Babeuf. This book was set in Century
Schoolbook by McGraw-Hill's Professional Book Group composition
unit.*

Printed and bound by R. R. Donnelley & Sons.

*I dedicate this book to my wife, Mona,
who was with me all the way.*

Contents

Preface   ix
Acknowledgments   x

Chapter 1. Basic Electrical Measurements and Instrumentation   1

1.1. Introduction   1
    1.1.1.   Basic Electrical Parameters   1
    1.1.2.   Time-Variant Voltage and Currents   2
    1.1.3.   Alternating Currents, Sine Waves   3
    1.1.4.   Electrical Energy Storage Elements' Reactance   4
    1.1.5.   Terminal Impedance and Resistance   5
1.2. Measurement Accuracy   5
    1.2.1.   Significant Figures of Measurements   7
    1.2.2.   Analysis and Arranging of Data   8
    1.2.3.   Logarithmic Representation, the Decibel   13
    1.2.4.   Electrical Measurement Standards, SI Units   15
    1.2.5.   Errors in Measurements   17
1.3. Basic Instruments   18
    1.3.1.   Voltmeters, Ammeters, Wattmeters   18
    1.3.2.   The Electronic Multimeter   24
1.4. The Oscilloscope   27
    1.4.1.   The Oscilloscope Display   27
    1.4.2.   Oscilloscope Improvements   32
1.5. Instruments with CRT Displays   36
    1.5.1.   The Television Waveform Monitor   36
    1.5.2.   The Spectrum Analyzer   38
    1.5.3.   The Time Domain Reflectometer (TDR)   41
    1.5.4.   The Network Analyzer   42
1.6. The LCD Display   42
    1.6.1.   LCD Screens and Bar Graph Displays   42

Chapter 2. Measurement of Electrical Quantities and Components   45

2.1. Electrical Parameters   45
    2.1.1.   Voltage   45
    2.1.2.   Measurement of Current   49
2.2. Resistance, Capacitance, Inductance   53

2.2.1. Resistance Measurement   54
2.2.2. Capacitors   57
2.2.3. Inductances   60
2.3. Impedance Measurements   62
2.3.1. Instrument Input/Output Impedance   63
2.3.2. High Frequency Effects of Instrument Cables   86

Chapter 3. Measurements at Audio Frequencies   71

3.1. The Audio Signal   71
3.1.1. Audio System Standards   72
3.1.2. Audio Power, Gain Measurements   74
3.1.3. Amplifier Distortion   76
3.1.4. Audio Noise Causes and Effects   81
3.2. Crosstalk   85
3.2.1. Stereo Separation   85
3.3. Home Audio Equipment   90
3.3.1. Audio Tape Recording   91
3.3.2. Phonograph Records   94
3.3.3. The Compact Disk Testing Method   96
3.4. Summary   97

Chapter 4. Measurements at RF Frequencies   99

4.1. RF Communications and Equipment   99
4.1.1. RF Electrical Measurements   99
4.1.2. Special Purpose RF Testing Equipment   103
4.2. RF Broadcast Equipment   108
4.2.1. FCC Rules for Broadcast   108
4.2.2. The Television Transmitter   111
4.2.3. FM Stereo Broadcast Service   132

Chapter 5. Microwave Systems Testing and Measurements 139

5.1. Line of Microwave Communications   141
5.1.1. The Reflex Klystron   141
5.1.2. The Magnetron   141
5.1.3. The Travelling Wave Tube   142
5.1.4. Solid-State Microwave Devices   143
5.1.5. Microwave Instruments   143
5.2. Microwave Communications Modulation   146
5.2.1. The Line-of-Site Microwave Link   147
5.3. Satellite Communications Systems   150
5.3.1. Satellite System Parameters and Instrumentation 151
5.3.2. End-to-End Testing   157
5.4. Radar Systems   161
5.4.1. Marine Radar Systems   162
5.4.2. Radar System Maintenance   166

Chapter 6. Cable Communication Systems   171

  6.1. Cable Systems   171
        6.1.1. The Twisted Wire Pair   171
        6.1.2. Telephone Line Equalization and Loading   178
        6.1.3. Coaxial Cables   178
  6.2. Fiber-Optic Cables   186
        6.2.1. Fiber-Optic Cable Systems   189
  6.3.. Cable System Testing   190
        6.3.1. Coaxial Cable Testing   190
  6.4. Fiber-Optic Cable Communications   208
        6.4.1. Fiber-Optic Cable Parameters   209
        6.4.2. Fiber-Optic Cable System Problems   210

Chapter 7. Digital Systems   213

  7.1. Digital Systems Development   213
        7.1.1. The Binary Numbering System   214
        7.1.2. The Digital Word or Byte   217
        7.1.3. Development of Digital Instruments and Controls   218
  7.2. Digital Test Equipment   226
        7.2.1. The Logic Probe, Logic Pulser   227
        7.2.2. The Logic Analyzer   228
        7.2.3. Digital Communications Testing Using BERT and
                Data Analyzers   230
        7.2.4. Software-Controlled Diagnostic Tests and Advanced
                Methods   232

Appendix A. The XYZ's of Using a Scope   235

  Introduction   235
  Part I.   Scopes, Controls, and Probes   235
            The Display System   238
            The Vertical System   241
            The Horizontal Sytem   248
            The Trigger System   254
            All about Probes   263
  Part II. Making Measurements   267
            Safety   270
            Getting Started   271
            Measurement Techniques   274
            Scope Performance   289
  Conclusion   292

Appendix B   Impedance Matching and Signal Level Adjustment   295

Bibliography   301

Index   305

Preface

Many practicing engineers and technicians take pretty much for granted manufacturers' published spcifications in brochures, catalogs, and other advertising media. Seldom is the question raised as to what tests or test procedures were used to verify a published specification. All too often problems occur when assembling and connecting equipment together to form a system. Special cables and interface systems may be needed to make the system function as planned. Many books on the subject of testing and measurements delve deep into the design philosophy of instruments and do not stress the ease of use and instrument accuracy that is important to the personnel using the instrument.

The first chapter discusses the basics of making electrical measurements and methods of analyzing the measurement results. Included is a review of the basic instruments and their development to present-day standards and configuration for a variety of uses.

Chapter 2 contains instrumentation and methodology for measurement of direct and alternating currents, voltages and power. Also, present-day methods of resistance, capacitance, and inductance measurements are studied. The importance of instrument input impedance loading and isolation is introduced.

The subject of audio system testing is contained in Chapter 3. The amplification recording and reproducing of audio signals for monaural and stereo methods is studied in sufficient detail. Audio signal levels distortions, crosstalk, and noise abnormalities are defined. Testing of the audio signal for these parameters is discussed and testing procedures are given. The special problems arising from the various audio recording media are given along with the associated testing methods for such signal degradation.

Chapter 4 contains a discussion of radiofrequency measurements as well as testing methods and procedures for the broadcast industry. Basic radiofrequency measurements of voltage, current, and power are introduced. The measurement of RF carrier frequency is discussed along with test on transmission lines for VSWR and return loss. Both AM and FM broadcast testing methods for carrier power, modulation, and frequency are included. Television video/audio broadcast testing is covered in detail in this chapter. Also included are FM stereo and the television stereo audio methods and tests.

Microwave communications systems and satellite communications system are covered in Chapter 5. The generation and measurement of microwave power are included. Impedance testing by SWR and return loss are discussed. Testing procedures for line-of-site microwave communication links as well as testing procedures for satellite up/down links are included. A study of radar systems is introduced, along with testing and maintenance procedures for a shipboard marine radar system.

Since a large part of worldwide communications systems operate as a wired or cable system, Chapter 6 is devoted to the measurement of various cable systems. The three main types of cables are covered in sufficient detail, beginning with metallic telephone cable to coaxial cable and fiber-optic cable. Techniques are covered for measuring the loss and continuity of such cable and loop resistance for the metallic cables. Fault finding, splicing, and special techniques are introduced. Examples of such systems along with some more general tests for telephone systems, cable television system, and local area networks are discussed.

Chapter 7 contains material on digital system methods and testing. The basics are included as a quick review and to introduce basic logic circuit testing. Digital displays and microprocessor controls are discussed, as are the various testing methods and instruments. Digital communication test methods and instrumentation are included along with examples of test procedures.

It was my intention in writing this work that its contents will act as a reference guide to aid in the establishment of proper and accurate testing procedures needed to verify performance specifications or aid in diagnostic testing.

Acknowledgments

Special thanks goes to the Tektronix Corporation of Beaverton, Oregon for their aid and support and to my good friend of nearly 40 years, Mr. Edward M. Vaughan, Customer Service Manager Television Division for Tektronix and his assistant Terri Boggs who kept me up to date. Also special thanks to Mr. Stephen Kowalchuk for his help in preparing the graphics.

Eugene R. Bartlett, P.E.

1

Basic Electrical Measurements and Instrumentation

1.1 Introduction

A wise person was once purported to say "To measure is to know." When one thinks about this quote, it makes a lot of sense. The two main reasons for making measurements on electronic operational equipment are either to test the systems to see if performance specifications have been met or to troubleshoot the equipment during repairs or routine maintenance. Such measurements may be as simple as measuring the input power, current, or voltage, or as complicated as applying various input signals and measuring the system's output response. Depending on the quantity of units to test, the instruments used may be simple manually operated types such as voltmeters, current meters, and/or single-signal generators, and so forth. Automatic or semi-automatic methods become cost effective when large quantities of units have to be tested. Many automated instruments are controlled either by a microprocessor or a computer. The parameters are selected and measured and the data gathered and analyzed by the computer or by a computer operator.

1.1.1 Basic Electrical Parameters

The basic electrical parameters common to all electronic equipment are the unit of electromotive force, the volt; the unit of current, the ampere; and the ratio of voltage to current, the resistance. When the voltage is measured in volts, and the current in amperes, the resistance is in the units called ohms. The unit of electric power for direct current circuits is the watt and is the product of voltage as measured in volts and the current as measured in amperes. Especially in communications circuits, power ratios are expressed in base 10 logarithmic form, and when multiplied by 10 the power ratio is expressed in decibels, abbreviated db.

1.1.2 Time-Variant Voltage and Currents

When the voltage, current, and power sources vary with time, then added electrical parameters are needed. Such time-variant sources are illustrated in Fig. 1.1. A parameter is needed to describe this time-variant phenomenon called the *frequency*. This parameter is a measure of how the signal repeats itself in the time period of one second, that is, the number of cycles per second. The time it takes for one cycle

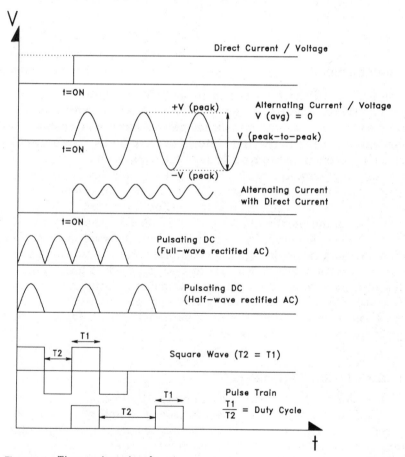

Figure 1.1 Time-variant signals.

to occur is defined as the *period*. An example of commercial electric alternating sine wave current is shown in Fig. 1.2.

1.1.3 Alternating Currents, Sine Waves

The frequency of alternating currents as well as time-variant current sources exhibiting periodicity may be expressed mathematically as: Frequency = $f = 1/p$, where p is the time for one complete cycle in seconds. Formerly the unit for frequency was cycles per second. Today the unit for frequency is the Hertz, abbreviated Hz. The unit is named after the German scientist Rudolph Heinrich Hertz. One hertz equals one cycle per second.

Example A sine wave voltage source at 1,000,000 Hz = 1×10^6 Hz or one megahertz (1 MHz). The time for one cycle is: $1 \times 10^6 = 1/p$ and $p = 1/(1 \times 10^6)$ s or one microsecond ($1\mu s$).

The manner of expressing the power, voltage, and current values for alternating currents as equivalent to the direct current values is by expressing these parameters as root mean square (rms) values. Transposing the letter R with the letter S (SMR) points to the method for calculating the RMS value. Therefore for the waveshape over one period, the amplitude value is squared, then the mean is calculated and finally the square root is taken. Mathematically, the rms value for current is expressed as

$$I_{\text{rms}} = \sqrt{\frac{1}{P}\int_0^P (f(t))^2\, dt}$$

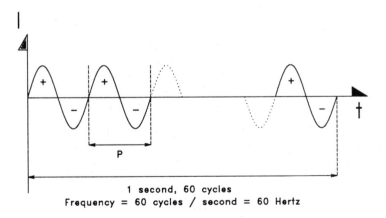

1 second, 60 cycles
Frequency = 60 cycles / second = 60 Hertz

P = Period = Time for 1 cycle = $\frac{1}{60}$ second

Figure 1.2 Time/period illustration for a sine wave.

where $f(t)$ is the mathematical expression for the waveshape. For a sine wave, $f(t) = I_{max} \sin 2\pi t/P$ which when calculated out using the calculus gives

$$I_{rms} = \frac{1}{\sqrt{2}} I_{max} = 0.0707 I_{max}$$

For waveshapes with vertical sides, one can calculate the rms value graphically as shown in Fig. 1.3.

1.1.4 Electrical Energy-Storage Elements' Reactance

When time-variant current and voltage sources are used as signal sources in many pieces of electrical and electronic equipment used today, the energy-storage elements of capacitance and inductance become very important. For direct current (dc) a capacitor would charge to the maximum value of the dc voltage. Also, when direct current was applied through an inductor coil, a steady magnetic field surrounded the inductor coil. For the alternating current (ac) condition, the opposition to the change in current was exhibited by the induction coil and expressed in ohms of inductive reactance. When an alternating volt-

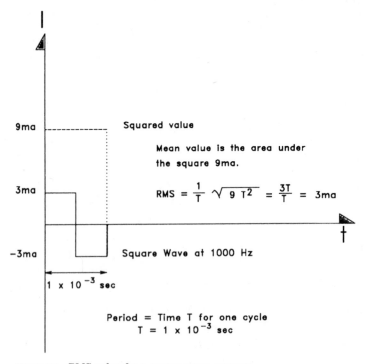

Figure 1.3 RMS value for a square wave current.

age source was applied to a capacitor, the capacitive reactance opposed the change in voltage. This reactance was also expressed as capacitive reactance in ohms. Essentially energy is stored in the magnetic field surrounding an inductor coil while energy is also stored in the capacitor's dielectric. The capacitive reactance of a capacitor reacts opposite to that of the inductive reactance of the inductor, which when placed in a series or a parallel configuration can cause series resonance (a short circuit) or parallel resonance (an open circuit). This short circuit (zero impedance) and open circuit (infinite impedance) act as pure resistances at resonance. This resonance condition takes place principally at a single frequency and is called the *resonant frequency*. The parallel or series circuit's impedance (opposition) to ac flow varies as the frequency of the ac source is changed from the resonance condition. The units for impedance and reactance, both inductive and capacitive, are in ohms because they oppose current flow. However, since they are energy-storage elements and charge and discharge energy, no power is consumed except a small amount in the resistance associated with the inductor coil or the capacitor's dielectric material. Most instruments used to measure impedance express the results in ohms, whether it is inductive, capacitive, or resistive.

1.1.5 Terminal Impedance and Resistance

All devices, whether instruments or devices to be measured, have terminal impedance, usually specified by the manufacturer. Caution should be exhibited when connecting instruments to devices, making sure that the instrument essentially does not change the operation of the device to be checked. Many instruments have their measurement terminals at a very high impedance and are as purely resistive as possible so as to not draw any current from the device being measured. This condition is often referred to as *system loading*. An example shown in Fig. 1.4 illustrates both conditions. This example is a difficult one because the circuit under test is at a very high impedance level. Typically many circuits are on the order of 1 to 10,000 ohms and most instruments have input impedances of 10 to 20 megohms. Thus the error is on the order of 0.1 percent.

1.2 Measurement Accuracy

The accuracy of a measurement of course, is extremely important. However, many measurements do not require extremely accurate results. Some measurements are only needed to verify that a circuit or device is operating, while others verify an important specification to the user of the device.

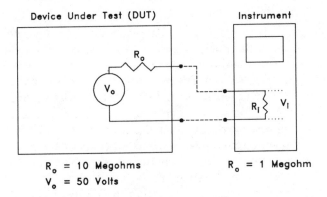

R$_o$ = 10 Megohms   R$_o$ = 1 Megohm
V$_o$ = 50 Volts

Normally, 50 Volts appears across the terminals at the DUT.
The Instrument indicates V$_I$.

$$V_I = \frac{R_I}{R_I + R_o} \times V_o = \frac{1 \times 10^6}{10 \times 10^6} \ 50 \quad \text{Volts}$$

$$= 5 \text{ Volts} \quad \text{(A gross error)}$$

Rule of thumb: R$_I$ should be 100 R$_o$.
When R$_I$ = 1000 Megohms then the error will be

$$V_I = \frac{1000 \times 10^6}{1010 \times 10^6} \ 50 = 49.5 \text{ Volts} \quad \text{(An error of 1\%)}$$

Figure 1.4   Instrument loading error.

The accuracy of a measurement is defined as the difference between the measured value and the true value (usually specified). The accuracy is often given as a ratio or a percentage of the true value. Mathematically, accuracy is expressed as

Accuracy = (measured value − true value)/true value

% Accuracy = [(measured value − true value)/true value] × 100

Unfortunately, in some cases the true value of a device is unknown, and establishing an accurate measured value by a properly calibrated instrument is the only means of properly assessing a device's operation. Therefore the known accuracy of the instrument making the measurement is important. The true value of a measured quantity is often a specified value given a system as a design goal. For example, the 115-V ac line voltage specified by the power utility company is a specified value which as most of us know varies by a few volts up or down.

Many analog instrument scales of the usual electromechanical type have small graduations that are difficult to line up with the indicating needle without the use of a mirrored scale to eliminate parallax or without the aid of a magnifying glass. Such instruments usually give an accuracy specification as plus or minus (±) some percentage of the full-scale reading. For example, if the full-scale value of a voltmeter is 1000 V dc with an accuracy of 1.5 percent, the reading could range from 1000 V ± 15 V. Since the 1015-V mark is past the 1000-V full-scale mark, i.e., maximum scale, the reading could be unknown above 1000 V. If the meter indication is −15 percent, then the instrument could be reading 985 V, which is on-scale. Also, if the instrument indicated 900 V on the 1000-V scale, then the true reading could range from 915 to 885 V. This 15-V number is referred to as the *range of doubt,* i.e., the possibility of error or range of possible error. If, for example, two such instrument readings have to be added together with two different instrument accuracy specifications, one simply has to add the magnitudes of the maximum errors for the final possible error. Proof of this is shown as follows:

Read #	Value, V	% Error	Min., V	Max., V	Error, V
$R_1 =$	100	± 2	98	102	± 2
$R_3 =$	150	± 1.0	149	151	± 1
Sum =	250	± 1.2	247	253	± 3
% Error =	(253 − 250)/250 = 3/250 = .012 = 1.2%				+ 2 + 1 = + 3
					− 2 − 1 = − 3

1.2.1 Significant Figures of Measurements

Since instrument readings are usually represented as a series of numbers, the practical number of significant digits must be considered. One must be practical in applying the needed accuracy to the type of measurement. After all, if it is desired to test whether a 15-V power supply is functioning properly, taking a measurement with a 4½-digit digital voltmeter and reading this value as 14.955 V seems ludicrous. A 3½-digit instrument will do as well. Just because an instrument (or calculator for that matter), gives more significant figures does not actually improve accuracy. Therefore, rounding of instrument readings to a proper number of significant figures is practical. When using a pocket calculator, we know that when an even number is divided by an odd number, a whole display of repeating digits results. Therefore some rounding is needed. A general rule is that the dividend or product value should not have more digits than the least accurate of the numbers yielding the value. In some cases possibly one more significant digit may be needed. Zeros to the immediate right of a decimal point are not counted as significant digits if the number is less than 1.

Example

0.0015   two significant digits

0.00   no significant digits

Also zeros immediately following the decimal point with no nonzero figures following are considered as significant for values greater than one, for example, 20.0 has three significant figures.

To round off numbers representing instrument readings, one must refer to the instrument accuracy. For example, a frequency counter with seven digits available and a published accuracy of 1 percent plus or minus the least-significant digit takes a reading of 151.5042 MHz. According to the instrument's accuracy, the true reading could be $151.5042 \times 0.01 = 1.515042 \pm 1$. This number should be rounded to the same number of decimal places as the reading. Therefore, for ± 1.515042 the maximum possible error is

$$153.0193 - 149.9891 = 3.0302$$

$$3.0302/151.5042 \times 100 = 2\% \text{ maximum possible error}$$

Caution must be observed when rounding numbers. The general rule for rounding numbers is if the number after the one wants to round is 5 or more, increase the number by one. For example, 3.59756 rounded to four significant figures is 3.598. To round to three significant figures, the number becomes 3.60, where zero is significant. Again, to round this number to two decimal places the number is 3.6.

1.2.2   Analysis and Arranging of Data

Analysis and arranging of the data gathered from instrument readings often clarify the behavior of the device being tested. Most students may recall the bell-shaped or normal distribution curve relating the probability of occurrence for various grades or test scores. The normal distribution curve is also referred to as the *Gaussian curve* after the German mathematician Karl Freidrich Gauss. An example of a distribution of class test scores forming such a gaussian curve is shown in Fig. 1.5. Only seven points form this curve for the seven grade values used. For instance, if the grades are given numerically, for every 5 percentage points from 50 to 110 percent, 13 points would result, thus making a smoother curve.

For another example, suppose 50 measurements of a precision resistor are taken using an ohmeter that can read to 0.1 ohm and the number of observed readings plotted versus the measured value results in the curve shown in Fig. 1.6. If an instrument is used that could read the resistance value to 0.01 ohm, then the 50 readings would be distributed by more points on the curve. The curve, however, will have

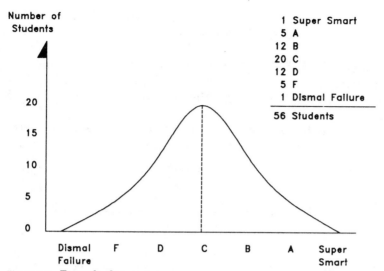

Figure 1.5   Example class test scores.

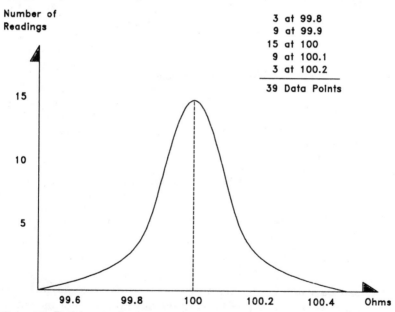

Figure 1.6   Resistance value test.

the same general shape. To relate measured data to the normal distribution curve, a number of terms should be defined.

1.2.2.1 Arithmetic mean. The arithmetic mean or average value of a series of instrument readings is simply the sum of the readings divided by the number of readings. Mathematically, $\bar{x}$ is the average reading, i.e., the most probable value.

x_1, x_2, x_3, x_n are the readings, where n = number of readings. Therefore,

$$\bar{x} = (x_1 + x_2 + x_3 + \cdots + x_n)/n.$$

Example

$$x_1 = 50.5 \text{ V}, x_2 = 49.3 \text{ V}, x_3 = 49.2 \text{ V}, x_4 = 50.3 \text{ V}, x_5 = 50.1 \text{ V}$$

$$\bar{x} = [50.5 + 49.3 + 49.2 + 50.3 + 50.1]/5 = 49.88 \text{ V}$$

1.2.2.2 Deviation from the mean value. The deviation or departure from the mean value is given as the difference between the reading and the mean value. Mathematically, let d = departure or deviation.

$$d_1 = x_1 - \bar{x} \text{ and } d_2 = x_2 - \bar{x}, \text{ etc.}$$

$$d_n = x_n - \bar{x}$$

Example   Using data from the previous example,

$d_1 = 50.5 - 49.88 = + 0.62$

$d_2 = - 49.88 + 49.3 = - 0.58$

$d_3 = - 49.88 + 49.2 = - 0.68$ adding this last column equals zero

$d_4 = 50.3 - 49.88 = + 0.42$

$d_5 = 50.1 - 49.88 = + 0.22$

1.2.2.3 Mean Deviation. The mean deviation or average deviation is more appropriate where one is looking for an indication of measurement precision. The mean deviation D is simply the sum of the deviation magnitudes divided by the number of measurements. Mathematically,

$$D = \frac{|d_1| + |d_2| + |d_3| + |d_4| + \cdots + |d_n|}{n}$$

For the above example, D may be calculated as

$$D = (0.62 + 0.58 + 0.68 + 0.42 + 0.22)/5 = 2.52/5 = 0.504 \text{ V}$$

1.2.2.4 Standard deviation. Statistical analysis of random errors defines the standard deviation as the square root of the sum of the squared deviations divided by the number of measurements. Mathematically, for n number of readings the standard deviation is given as

$$\sigma = \sqrt{\frac{d_1^2 + d_2^2 + d_3^2 + \cdots + d_n^2}{n}}$$

However, for n not to be an infinite number of observations, $n - 1$ will suffice for a finite number.

$$\sigma = \sqrt{\frac{d_1^2 + d_2^2 + d_3^2 + \cdots + d_n^2}{n-1}}$$

1.2.2.5 Error probability, probable error. The area under the Gaussian curve for various multiples of σ is important in assessing the error probability. This is shown in Fig. 1.7. However, the curve of the data

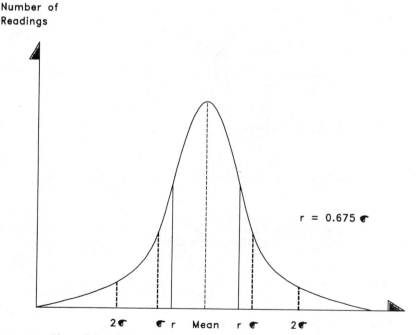

Figure 1.7 Normal distribution curve.

points, that is, the instrument readings, do not have to be plotted to do a statistical analysis of normally distributed values. The above mathematical formulas can be used. One other value shown in Fig. 1.7 is the area of probable error γ, which is found to be $\pm 0.675\sigma$ or about two-thirds of σ. The magnitude of γ is the probable error in 50 percent of the number of measurements. If only one measurement is taken, the probable error is $\pm 0.675\sigma$. An example will illustrate the principles presented here.

Example A instrument has measured a capacitor for the following values of capacitance [in picofarads (pF)] 102, 105, 110, 112, 107, 108. The average capacitance value (C_{avg}) can simply be found as

$$C_{avg} = [102 + 105 + 110 + 112 + 107 + 108]/6 = 644/6 = 107.333$$

The average deviation can be found by

$$D_{avg} = \frac{|d_1| + |d_2| + |d_3| + |d_4| + |d_5| + \cdots + |d_n|}{n}$$

$$D_{avg} = |107.33 - 102| + |107.33 - 105| + |110 - 107.33| +$$

$$\frac{|112 - 107.33| + |107.33 - 107| + |108 - 107.33|}{6}$$

$$= (5.33 + 2.33 + 2.67 + 4.67 + 0.33 + 0.67)/6$$

$$= 16/6 = 2.67 \text{ pF}$$

which says the measured value will vary on the average by this amount. The standard deviation can be calculated by

$$\sigma = \sqrt{\frac{d_1^2 + d_2^2 + d_3^2 + \cdots + d_n^2}{n - 1}}$$

$$\sigma = \sqrt{\frac{5.33^2 + 2.33^2 + 2.67^2 + 4.67^2 + 0.33^2 + 0.67^2}{6 - 1}}$$

$$\sigma = \sqrt{\frac{28.41 + 5.43 + 7.13 + 21.81 + 0.11 + 0.45}{5}} = \sqrt{\frac{63.34}{5}} = 3.56$$

The probable error may be calculated as $0.675\sigma = (0.675)(3.56) = 2.4$ pF. This means that one-half of the time the error will most likely be 2.4 pF. Notice that the probable error is numerically closer to the average deviation than the standard deviation. However, statisticians believe the standard deviation (1σ) is more appropriate when analyzing the dispersion of measured values. Simply stated, small variations from the mean value are much more probable ($\pm 1\sigma$) than larger deviations, $\pm 2\sigma$, $\pm 3\sigma$, etc. It must be remembered that the larger the number of samples, the smoother the normal distribution curve becomes and the mean value represents the most probable value.

1.2.3 Logarithmic representation, the decibel

Before discussing specific instruments and getting involved with equipment, logarithmic representation of electrical parameters such as power, voltage, and current, should be thoroughly understood. The decibel is nearly universal in describing the power ratio in the logarithm form. The decibel simply means $\frac{1}{10}$ bel. The unit is named after Alexander Graham Bell who was a teacher and authority on the deaf as well as the inventor of the telephone. Bell discovered that the human ear responded to logarithmic changes in acoustic power levels. The invention of the telephone was the product of Bell's work in attempting to aid the deaf. By definition, the bel is simply defined as

$$\log P_1/P_2$$

where P_1 and P_2 are in the same units, that is, watts, kilowatts, milliwatts, etc. Since this is a large unit, $\frac{1}{10}$ bel (decibel, dB) becomes more practical and is defined as

$$dB = 10 \log P_1/P_2.$$

Most people working in the electrical and electronic industries have had sufficient mathematics to handle logarithms, logarithmic graph paper, and to make calculations of products by adding the logarithms and finding the antilog or inverse log. Numbers represented in decibel form can be represented as a sum of decibels for a product of power ratios and a difference for a division of power ratios. Hence the signal-to-noise ratio can be represented as the S/N ratio and is the decibel value of the signal power level minus the decibel value of the noise level. Since the decibel represents a power ratio, in order to represent a power-level value, a zero decibel value has to be determined. The level dBm means a decibel value above or below 0 dBm, where 0 dBm is referred to 1mW, hence dBm.

$$\text{No. dBm} = 10 \log P/1 \text{ mW}$$

Example   A radio transmitter has an output power of 5 W. What is this value in dBm?

$$\text{No. dBm} = 10 \log 5000/1 = 10 \times 3.70 = 37 \text{ dBm}$$

Example   A radio receiver receives a signal at 10 microwatts (μW). What is this value in dBm?

$$\text{No. dBm} = 10 \log 1 \times 10^{-5}/1 = 10(-5) = -50 \text{ dBm}$$

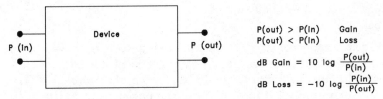

Figure 1.8 Defining the decibel.

Most people reasonably versed in mathematics know that the logarithm of a number is merely the power the base is raised to, in this case, -5. This calculation can be easily done using any of the reasonably priced scientific pocket calculators.

The power ratio discussed is most often the output power divided by the input power, which represents the power gain or loss of a device. An example is shown in Fig. 1.8. Since the logarithm of a number less than 1 cannot be taken because the logarithm of 1 is 0, the larger power has to appear in the numerator, and if it is the input power indicating a loss, the decibel value carries a negative sign.

Other values of power levels used today are shown in Table 1.1 along with the reference level.

The dBmV parameter is used mainly in the cable television and LAN industry (local area network). Since voltages at the microvolt and millivolt levels are used in radio and television work and since impedance matching is so important, reference levels based mainly on signal voltages are used. Power is related to voltage by

$$P = V^2/R \quad \text{so } P_1 = V_1^2/R_1$$

and

$$P_2 = V_2^2/R_2 \quad \text{dB} = 10 \log P_1/P_2$$

for

$$R_1 = R_2 \quad \text{dB} = 10 \log V_1^2/V_2^2.$$

Recall the identity $\log x^n = n \log x$.

$$\text{dB} = 20 \log V_1/V_2 \quad V_2 = 1 \text{ mV} \quad \text{dBmV} = 20 \log V_1/1 \, mV$$

TABLE 1.1

dB	Reference Level
dBm	1 mW
dBu	1 μW
dBw	1 W
dBmV	1 mV (across 75 ohms)

One conversion often used is to refer to some dBmV value as a dBm value. To convert the equivalent value of 0 dBm to dBmV, calculate the power consumed by the 75-ohm resistor with 1 mV across it and then calculate the dB value for the difference. This is illustrated in Fig. 1.9. Therefore if an instrument reads a value across a 75-ohm system as -30 dBm then this value corresponds to $-30 + 48.75 = -18.75$ dBmV.

It must be remembered that the decibel is based on the power ratio, and even if 20 times the logarithm of the voltage or current ratio is used, it is a power ratio and the circuit impedances for the voltage or current ratio have to be equal. Since the decibel unit is used so extensively in the electronics industry, many instruments have scales or readouts in dB. Two more examples are shown in Fig. 1.10a and b and illustrate the extensive use of the dB.

1.2.4 Electrical measurement standards, SI units

For electrical and electronic equipment to function in the world today from country to country, standards of measures and procedures have to be set. Most of us have struggled with various systems of measure such as the English system, cGS (centimeter, gram, second), mKS (meter, kilogram, second), etc., and the need to convert back and forth. In 1954, by international convention, the ampere was added to the mKS system creating the mKSA system, and the Kelvin and candela were added as units for temperature and luminosity to form the SI system of units. This system was implemented in 1960 and is the internationally accepted system of units used today. The study of physics relates the units of length (meter), mass (kilogram), and time (seconds) to force. The force in turn is related to electric charges and magnetic poles by strict mathematical procedures. Since electric charges

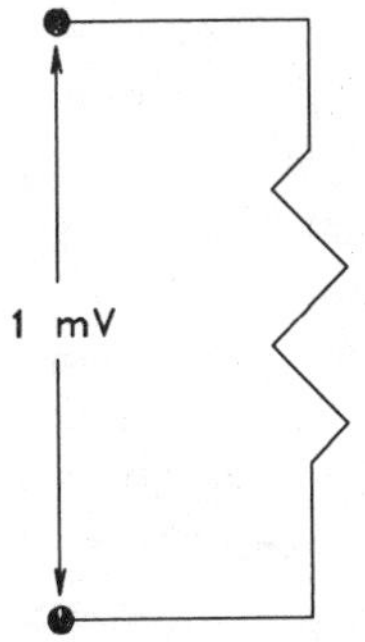

$$P = \frac{V^2}{R} = \frac{(1 \times 10^{-3})^2}{75}$$

$$= 0.0133 \times 10^{-6} \text{ w} = 13.3 \times 10^{-3} \text{ mw}$$

$$dB = 10 \log \frac{1 \text{ mw}}{13.3 \times 10^{-6}} = 10 \log 0.075 \times 10^{6}$$

$$= 48.75 \text{ dB}$$

Figure 1.9 The dBm to dBmV.

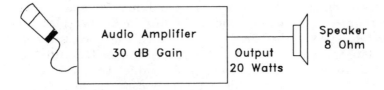

Find the input power for full output power

20 w corresponds to a dB level based on 1 watt.

dBw = 10 log 20 = 13 dBw

13 − 30 = −17 dBw Input power

−17 = 10 log P_1

$P_1 = 10^{-1.7} = 0.02$ w = 20 mw

(a)

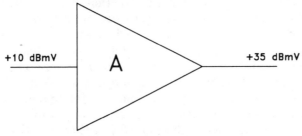

Find amplification A In dB.

35 − 10 = 25 dB Gain

(b)

Figure 1.10   Using the decibel.

are related to electric potential differences (volt) and magnetic force is related to electric current (ampere), the electrical parameters are specified. Also, since the frequency is related to wavelength through the speed of light, the unit of length and second are brought into play again. The SI units of time, temperature, mass, and distance represent primary standards on which other units are based. Seconds, Kelvin, kilograms, and meters represent these quantities. In the United States the National Bureau of Standards Laboratory in the State of Colorado maintains working standards to which industry equipment can be calibrated. In most major cities in the United States commercially operated calibration laboratories offer calibration services to the local industries. Such local calibration and repair facilities offer calibration services for resistance, capacitances, inductance voltage, current (both dc and ac), and frequency to traceable NBS standards. Also many calibration laboratories offer repair facilities for a variety of electronic test equipment. Of course most of the manufacturers of test equipment offer repair and calibration services for their own equipment, and many of these manufacturers maintain repair and calibration facilities in major cities or areas of industrial concentration. Repaired and calibrated equipment should have a written report of the calibration accuracy and the traceable standards used accompanying the returned test equipment.

Some major manufacturers of electronic systems and equipment have their own in-house calibration laboratory which maintains traceability to the National Bureau of Standards. This can be a cost-effective and time-saving method in keeping properly calibrated test equipment in use.

1.2.5   Errors in measurements

The problem often arises during a testing procedure that an error in measurement may be present. If an error is suspected, the question arises as to when the error took place and how many manufactured goods might be affected. A technique often found useful is a comparison of instrument readings of a common parameter, that is, a cross-check of instruments. If there is a difference in readings beyond the published accuracies of the instruments, then an error has occurred. The next thing to do is find out which instrument is in error. A case in point, if ten like instruments are cross-tested with each other and nine are within the published accuracy tolerance, then it could be safely assumed that the tenth one is in error. If this procedure is followed several times a day, say for each manufacturing shift, then when a measuring instrument fails or loses its accuracy, fewer of the manufactured products will be affected. In many cases instrument manufacturers are building in to their equipment some means of internal calibration. For example, the Tektronix Corporation's 7L12, 7L14 Spectrum Analyzer contained a built-in calibrator of sufficient accuracy. Also if a 7K11 plug-in amplifier was used, it too had an internal calibrator. Both of these calibrators could be cross checked, thus leaving little doubt that the instrument was functioning properly. Many oscilloscope instruments contain a reasonably accurate voltage standard to calibrate the vertical scale. More will be said about this later during the sections on specific equipment.

1.3 Basic Instruments

1.3.1 Voltmeters, Ammeters, Wattmeters

Even today, many voltmeters, ammeters, wattmeters, etc., use an electromechanical moving-coil meter called a D'Arsonval meter as the measuring device. An indicating needle covering about one-third of a circle (120°) that moves across a scale graduated in volts, amps, etc., is the basic indicating instrument. If the indicating needle has a length of 1½ in for a 3 in diameter meter, then the arc of 120° corresponding to one-third of a circle gives a meter scale length of ⅓ × π × d = ⅓ × 3.14 × 3 = 3.14 in. Now if the needle width is 0.015 in, then the maximum number of scale graduations with equal spaces in between can be calculated by: ½ × 3.14 in ÷ 0.015 in/mark = 105. Therefore in round numbers an easily read scale for a 3-in-diameter meter will have 100 marks and 99 spaces. So if this meter is read to the nearest mark, the accuracy will be about 1 percent or one mark in 100. This is about the best accuracy obtainable by such an indicator. Large meters with more scale length, thus containing more graduations, will give greater accuracy. To make reading the needle scale easier, a mirror strip is placed on the meter face usually just above or below the scale graduations. This improves the accuracy somewhat by eliminating parallax between the needle and the scale. About ½ percent of full scale is about the best accuracy obtainable with the electromechanical meter movements. It should be obvious that the internal circuit elements need not be much more accurate than the meter scale.

1.3.1.1 Electromechanical meters. The electromechanical meters belong to what is known as the analog domain. Essentially these meters have either a moving coil (dynamometer) or a moving vane that carries a pointer (indicating needle) across a scale. Figure 1.11 shows the moving-type meter. This type of a meter is essentially a dc motor that is allowed to turn only about one-third of a circle. The needle deflection is proportional to the current through the meter coil. Therefore to measure voltage current and resistance, internal circuitry had to be added to the basic meter movement. Instruments using electromechanical meter movements were more delicate, since the moving members were usually watch-jewel-mounted and the restraining coils were small and thin. Also the copper wire on the coil was very fine so as not to add too much mass to the moving coil. The resistance of many moving coil meters were on the order of 25 to 100 ohms and a specification for meter sensitivity was given in ohms per volt. The ohms-per-volt rating of a meter's sensitivity is the reciprocal of the meter movement's full-scale current.

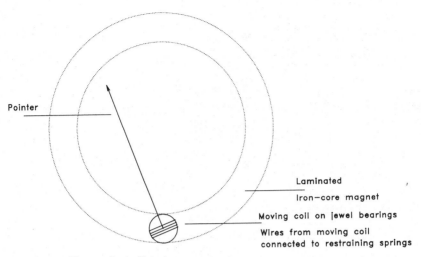

Figure 1.11   The moving-coil meter movement.

Example   A meter movement has a full-scale current of 1 mA. The meter sensitivity in ohms per volt is

$$S = 1/(1 \times 10^{-3})A = 1 \times 10^3 \text{ohms/V or } 1000 \text{ ohms/V}$$

For a meter with a 50-μA movement:

$$S = 1/(50 \times 10^{-6}) = .02 \times 10^6 = 20,000 \text{ ohms/V}$$

Recall $I = V/R$ = volts/ohm so $1/I$ = ohms/V

It stands to reason, the less current required for full-scale deflection, the more sensitive the meter.

1.3.1.2 Meter ranges. Adding series resistors will increase the voltage range of a meter movement, and placing resistors across the meter movement will increase the current rating. Figure 1.12a and b demonstrates the process using a 1-mA full-scale meter movement with a 25-ohm coil resistance. Figure 1.12a makes a 0- to 100-V voltmeter and 1.12b a 1-A meter. It must be remembered that electromechanical meter movements draw current from the circuit the meter is measuring, which may cause the circuit to malfunction and in turn cause the instrument to give incorrect results. An example of meter loading using the voltmeter in Fig. 1.12a causing incorrect results is shown in Fig. 1.13.

1.3.1.3 Meter resistance, circuit loading. If a meter movement has a sensitivity of 20,000 ohms/V, then 50 μA will be the full-scale current. Meters of this type have coils with more turns of wire and hence finergauge wire, thus causing higher resistance. Consider a meter move-

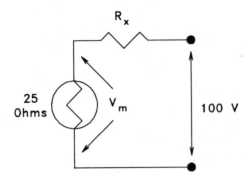

$$V_m = (1 \times 10^{-3})(25) = 25 \text{ mV}$$

$$I R_x = 100 - 0.025 = 99.975 = (\text{approximately}) \ 100 \text{ V}$$

$$R_x = \frac{100}{1 \text{ mA}} = 100 \text{ kOhm}$$

NOTE: 100 kOhm / 100 V = 1 kOhm / V

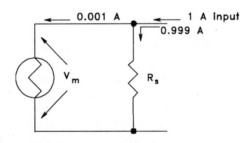

$$V_m @ \text{Full Scale} = 25 \text{ mV}$$

$$R_s = \frac{25 \text{ mV}}{0.999 \text{ A}} = \frac{25 \times 10^{-3}}{1 \ (\text{approx.})}$$

$$= 0.025 \text{ Ohm}$$

Figure 1.12   The voltmeter: (a) current meter; (b)

ment with a 200-ohm coil shown in Fig. 1.14. The increased sensitivity has decreased the effects of the meter on the circuit, with a large improvement in measurement accuracy.

Ideally a voltmeter should draw no current from the circuit being measured, that is, it should be an open circuit. Also, an ammeter should present nearly a short circuit, thus ensuring all the current

Error Caused by Meter Loading

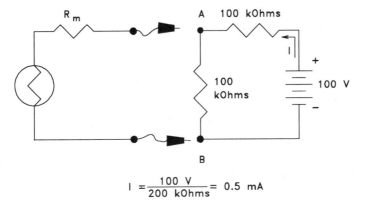

$$I = \frac{100 \text{ V}}{200 \text{ kOhms}} = 0.5 \text{ mA}$$

Connecting meter across a–b places meter's 100 kOhm resistor
in parallel. Therefore, the meter reads $\frac{50 \text{ K}}{150 \text{ K}}$x 100 V = 33.3 V.

This causes an error of $\frac{50 - 33.3}{50} = \frac{16.7}{50}$ = 33% Error.

Figure 1.13 Error caused by meter loading.

flows through the meter circuit and hence is measured. Voltmeters
are always placed across a pair of circuit terminals and ammeters are
placed in series.

1.3.1.4 Electromechanical watt meter. The moving-coil dynamometer-
type meter movement can be configured with two coils, one a voltage
coil, which is a moving coil connected to the pointer, the other a stationary field coil. The magnetic flux resulting from the magnetic fields
produced by both coils influences the movement of the moving coil
with the attached pointer. The connections for a wattmeter are shown
in Fig. 1.15. As the source voltage increases, the current drawn will
increase, in turn increasing the needle deflection proportional to the
true power drawn by the load. This type of meter is useful for both dc
and ac up to a few hundred hertz. The wattmeter connection often labeled ± is the connection where the potential (voltage) coil and the
current (field) coil is common. By adding a shunting resistor across
the current coil and adding a series resistor to the potential coil, the
voltage and current ranges can be controlled and hence the wattage
rating. Caution must be observed when specifying a wattmeter for a
particular application. The voltage may be too high for the potential
coil's rating even though the current coil can pass the expected load
current and not exceed the full-scale wattage rating.

Error Decreased by More Sensitive Meter

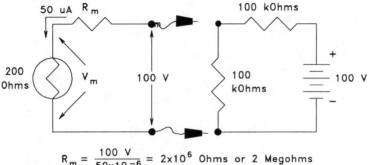

$$R_m = \frac{100 \text{ V}}{50 \times 10^{-6}} = 2 \times 10^6 \text{ Ohms or 2 Megohms}$$

Connecting across a and b places R across 100 kOhm resistor

$$\text{Parallel combination} = \frac{2 \times 10^6 \text{ x } 1 \times 10^5}{2.1 \times 10^6} = \frac{2 \times 10^{11}}{2.1 \times 10^6} = 95 \text{ kOhms}$$

$$\text{Voltage across a–b is } V_{ab} = \frac{95K}{195K} \times 100 = 48.7V$$

$$\text{Now, \% Error} = \frac{50 - 48.7}{50} \times 100 = 2.6\% \text{ instead of 33\%}$$

Figure 1.14   Error decreased by more sensitive meter.

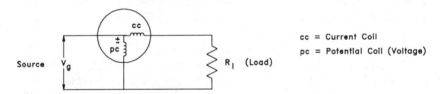

cc = Current Coil
pc = Potential Coil (Voltage)

Figure 1.15   Wattmeter connections.

1.3.1.5 Meter ac, dc operation. Since voltmeters and ammeters of the moving coil type cannot respond to alternating current, adding a diode (rectifier) in series with the meter coil will convert the applied alternating current or voltage to a pulsating direct current. The meter pointer will respond to the average of the pulsating voltage or current. A switch can be placed to bypass the diode for dc operation. Such a meter will have one scale for dc and one for ac operation. The scale for ac operation will be calibrated for the effective (rms) value for sine wave operation only.

1.3.1.6 Measurement of resistance, the ohmmeter. To measure resistance in a self-contained instrument, some built-in power has to be applied to the resistance being measured where the meter movement essentially measures the voltage drop across the unknown resistance. A diagram of the simple ohmmeter is shown in Fig. 1.16. First the test probes are shorted (connected) together and variable resistor R_2 is adjusted for a full-scale deflection. Essentially 1½ V appears across terminals a-b. Inserting a resistor between the probes will cause the meter movement to decrease. Therefore zero ohms corresponds to full-scale deflection and deflection decreases as resistance increases, thus the meter scale is opposite that of voltage and current magnitude. The deflection can be calculated by using:

$$\% \text{ Deflection} = R_0/R_o + R_x \times 100$$

$R_0 = R_1 + R_2$ and R_x = resistance to be measured.

It is assumed that the meter's coil resistance R_m is negligible compared to $R_1 + R_2$.

Example

$$R_o = R_1 + R_2 = 5000 \text{ ohms} \quad R_x = 2500 \text{ ohms}$$

$$\% \text{ Deflection} = 5000/7500 \times 100 = 2/3 \times 100 = .667 = 66.7\%$$

or two-thirds of the scale.

As the value of the unknown resistance increases, the deflection gets less and the scale markings get bunched up near zero deflection.

Ohmmeter Circuit

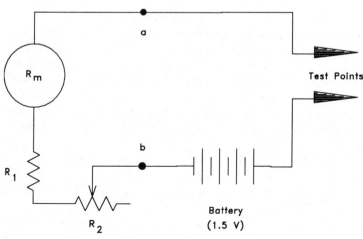

Figure 1.16 Ohmmeter circuit.

Therefore, increasing the battery voltage and switching resistance between terminals *a-b* can allow the meter ranges to be changed.

1.3.1.7 The Multimeter. Changing the position of the meter movement in the circuit configuration by means of a selector switch can allow the meter to be changed from a voltmeter, ammeter, or ohmmeter. Also, by selecting values of series and shunt resistors, the voltmeter, current meter, and ohmmeter ranges can be selected. Such a meter is appropriately called a multimeter and many are still in use today. When selecting multimeters of this type for a particular application, it must be remembered that the meter sensitivity is an important specification because in the voltmeter configuration less current is required from the measured circuit, and for the current meter the shunting resistor can be a lower value to create a condition nearer to a short circuit. Also, more sensitive meters can in general read much larger resistances. Naturally an instrument with more sensitive meter movement is more expensive and delicate, requiring caution in use and handling.

1.3.1.8 The high-impedance voltmeter. To minimize the circuit-loading effects due to a voltmeter's resistance, a unity-gain amplifier with an extremely high input impedance was used to isolate the meter resistance. Before solid-state devices were available, a vacuum tube amplifier was used. Hence a meter of this type was called a vacuum tube voltmeter or simply a VTVM. This type of instrument was also adapted to measure resistance as well. Some of these instruments are still in use today.

1.3.2 The electronic multimeter

The modern-day multimeter is often of the digital type. The first type of digital meter made the measurement by analog methods, the display being digital. Now essentially the measuring circuit is on a single chip which converts the measured parameter to a digital number which is scaled and converted to a digital LED (light-emitting diode) or LCD (liquid crystal display) display. Some chips have a computer circuit that can calculate the true rms value of the voltage or current measurement for nearly any waveshape and a wide frequency range. Such instruments also have autoranging capabilities and some form of input protection against overload. Most DVMs (digital multimeters) have a high input resistance for voltage measurements and the ability to measure small currents and a wide resistance range. The specifications for a typical DVM are shown in Table 1.2.

The method of conversion to digital numbers is different in some

TABLE 1.2 Typical DVM specifications

Parameter	Range	Accuracy
DCV	100–1000 V	± 0.5% of reading +1 digit
ACV	100–750 V	± 1.25% of reading +4 digits
DC current	200A–20A	± 1% of reading +1 digit
AC current	200A–20A	± 1.5% of reading +3 digits
Resistance	200 ohms–2000 Mohms	1% of reading to 2 Mohms
		3% of reading to 2000 Mohms

DC volts input impedance 10 Mohms
AC volts input impedance 10 Mohms shunted by 100 pF at frequencies of 40–400 Hz
(750-V range), 40–1 kHz, 750 V

meters. Basically, one of four methods is used: (1) ramp type, (2) integrating type, (3) continuous balance type, and (4) successive approximation type. The ramp-type converter essentially changes the voltage level to a time interval, and during the time interval clock pulses are counted. The count represents the voltage value. The integrator type is a variation on the ramp type which generates a staircase ramp which then generates a time interval to gate an oscillator. Lower-cost DVMs use a successive approximation type of A-D (analog-to-digital) converter. Through digital logic a number is generated and converted to an analog voltage which is compared to the input voltage through a comparator circuit. From a practical standpoint, the main criteria for choosing an instrument is by its published specifications and guaranteed accuracy. Some of the methods of converting to the digital domain determine speed of measurement, which may or may not be important. When making many repeated measurements, speed of measurement may become significant.

1.3.2.1 Computer-controlled multimeter. Many of the laboratory-type or bench DVMs have the measured value appearing at a connector on the rear apron of the instrument as a digital number either as a standard RS232 port or the IEEE 488 bus. A digital printer can be connected to this port or a PC (personal computer). The printer can provide a printed record of the measurement, whereas the computer, operating with one of the many data-handling programs, can sort, arrange, or manipulate the measured information and store it on a floppy disk for future use.

1.3.2.2 Multimeter accuracies. The accuracy of many meters is specified as a tolerance of the measured parameter plus or minus the LSD (least-significant digit). Some DVMs limit the display to 3½ digits, while instruments with a higher accuracy specification and measuring range will have 4½ to 5½ digits. The 1/2 digit simply means the MSD (most-significant digit) is a 0 (zero) or 1 (one). A four-digit instrument will have its MSD as 0 through 9. Recalling the fact that an electromechanical type of multimeter gives the accuracy as a percent of full-scale reading, while the digital-type meter gives the accuracy in percent of the reading plus or minus a number of digits, a comparison of accuracy is simple. The following example makes this comparison.

Example   A dynamometer (D'Arsonval-type) measures 50 V on the 100 V full-scale range with an accuracy of 3 percent of full scale. Accuracy is $\pm 0.03 \times 100 = \pm 3$ V. Reading could be 47 V or 53 V with an accuracy of $3/50 \times 100 = 6\%$ of reading.

The digital voltmeter making the measurement on its 100-V range with an accuracy of ½ percent of reading ±1 digit measured voltage will read 49.75 or 50.25 or 49.74 or 50.26. Even if the meter had a 3 percent of reading accuracy instead of 3 percent of full-scale accuracy, the difference in accuracy of reading is $6 - 3 = 3$ percent difference. So much for specsmanship.

1.3.2.3 Modern-day multimeters.

Multimeters are in wide use today as they have in the past. In general the digital-type instruments are more accurate and are rugged, and hence stand up under the rigors of travel and portable use. Also the batteries require more attention than the electromechanical meters, which only require a battery for the ohmmeter section. The digital meter uses the battery to perform all measurements and hence is all electronic. Some meters have rechargeable batteries with built-in chargers while others require such batteries as simple dry cells, alkaline cells, and/or mercury batteries. Many meters have a built-in battery test feature which enables the battery condition to be tested before making any measurements, thus avoiding erroneous measurements. The digital-type display generally eliminates human reading errors and hence improves measurement accuracy. The digital meter displays are either LED or LCD type. The LED type is more easily seen but causes more battery drain. The LCD type has dark figures against a shaded background and can be backlit to improve contrast. Most multimeters of either type are used mainly in troubleshooting and maintenance work. Almost every field technician or field engineer carries a multimeter usually of the digital type in their tool and equipment luggage. One feature the electromechanical meter movement had that many people regarded as important was its ability to display slowly or flickering values on the meter face. Digital meters would show this as the less-significant digits flashing by. Manufacturers of digital multimeters solved this problem by adding a bar graph feature to its digital display. Now they had the best of both worlds. Audio tones have been added to many digital-type multimeters to aid in the verification of circuit continuity. To make more use of the more sophisticated LCD displays features such as a function generator, frequency counter, and inductance-capacitance testers were added, making the DVM a virtual component tester as well as a simple multimeter. Many field technical people still prefer the DVM and separate units for other testing.

1.4   The Oscilloscope

The modern-day oscilloscope is probably still one of the most powerful instruments in use today. The old simple oscilloscope presented a mere amplitude versus time presentation of a voltage waveform on a cathode-ray tube display. Modern-day oscilloscopes have sophisticated synchronizing circuits, accurate calibrated vertical scales with wide signal bandwidth, fast accurate linear sweep circuits, as well as the ability to generate several traces for signal comparison and blanking systems to accentuate portions of a waveform. Much to everyone's satisfaction is the fact that the costs have decreased to a point where a decent oscilloscope is within most everyone's budget. Many of the oscilloscopes today even offer portability, with battery powering options available. In the past few years many of the more sophisticated oscilloscopes operate in the digital domain. These instruments present the usual oscilloscope cathode-ray tube display but in essence the display is produced by a digital memory containing the waveform to be measured. Some instruments have the ability to store waveforms in solid-state memory banks for later reference or analysis by a computer system. Also available are oscilloscopes with the ability to be setup and adjusted under a PC program and data extracted as well. Many of the manual oscilloscope controls operate digitally therefore the computer control setup is really quite simple.

1.4.1   The oscilloscope display

The cathode-ray tube display still appears in many types of instruments even today. However, the appearance of the LCD screen displays are seen more and more in modern-day instruments. These displays do not need high voltages to operate, require less circuitry, and are low in cost. However, speed of real-time measurements presents a problem to the LCD type of screen display.

1.4.1.1   The cathode-ray tube.   The basic cathode-ray tube (CRT) is shown in Fig. 1.17. A concentration of electrons is formed by the cathode, which is heated by the filament. These electrons are formed into

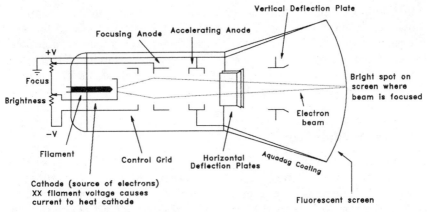

Figure 1.17 Basic cathode-ray tube.

a beam by the focusing and accelerating anodes. With no voltage between the control grid and cathode and deflecting plates the beam will hit the phosphored front screen causing a spot of light. Making the control grid more negative with respect to the cathode will cause the bright spot to diminish. Placing voltages on the horizontal deflecting plates will cause the spot to move laterally, and if the vertical plates are charged the spot will move vertically. It should be evident that by controlling the electrode voltages on the CRT one can change the brightness of the beam and place it anywhere on the screen face. For oscilloscope operation a sweep voltage is placed on the horizontal deflection plates, causing the beam to start on the left center and move to the right. For a very short instance the beam is turned off by the control grid, while the voltages on the plates reverse polarity moving the beam back to left center, ready to form another visible left-to-right sweep. This is shown in Fig. 1.18. Now the voltage to be displayed causes the voltage on the vertical deflection plates to change, moving the beam in the vertical direction. By controlling the sweep rate to correspond to the period of the vertical axis signal a complete cycle of the amplitude changes of the vertical signal will be displayed. This is shown for one cycle of a sine wave in Fig. 1.19.

1.4.1.2 Development of the oscilloscope. Earlier oscilloscopes had few controls available, which were mainly the vertical input level control and the horizontal sweep rate control, the horizontal and vertical positioning controls, the intensity and focus control, and last but not least, the power switch. It indeed was a problem to obtain a one or two cycle display of a waveform to stand still on the screen for analysis. Some form of signal synchronization was sorely needed. A triggering

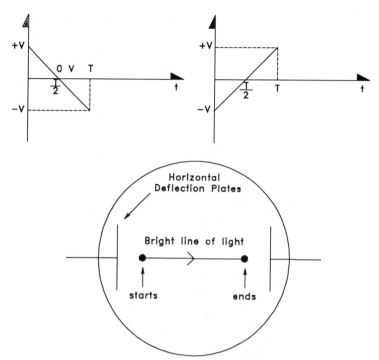

Figure 1.18   Horizontal beam sweeping, end view of cathode-ray tube.

circuit was added to examine the input signal level and start the sweep circuits to form the display. The level point which started the sweep circuits was selectable by a front panel control. This aided the operator to display a desired portion of the input signal. The trigger circuit was further developed to select a positive or negative going portion of the input signal for triggering. This extended the usefulness of the oscilloscope to measure signal waveforms.

1.4.1.3 Oscilloscope measurements. With an accurate sweep time, scale measurements of frequency were possible. Voltage level at various points of the input signal were possible when the vertical scale was calibrated accurately. On the screen of the CRT a plastic scale (graticule) was placed with vertical lines and horizontal lines engraved on it. By calibrating the vertical scale in volts per division (usually volts per centimeter) and the horizontal scale in time per division (usually seconds, milliseconds, and microseconds per centimeter). Reasonably accurate measurements of voltage and frequency could be made. Addition of a built-in voltage calibrator circuit aided making more accurate amplitude measurements. Such a screen using

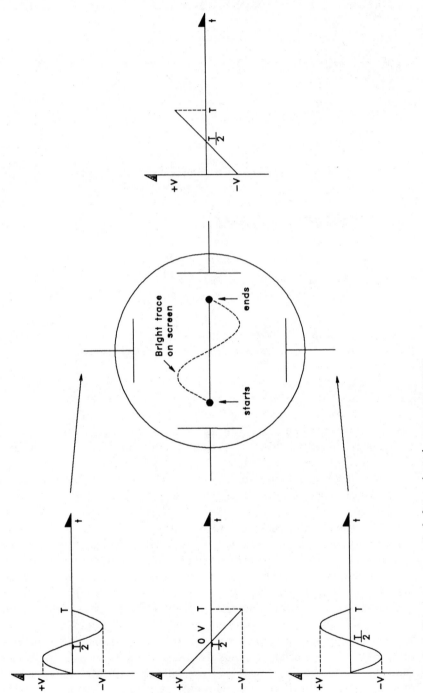

Figure 1.19 Sine wave on vertical plates gives sine wave on screen.

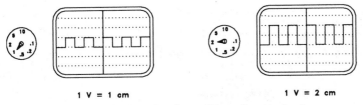

1 V = 1 cm   1 V = 2 cm

Figure 1.20   Calibrated vertical scale, oscilloscope vertical axis.

a built-in voltage calibrator is shown in Fig. 1.20. Figure 1.21 shows a simple set of oscilloscope synchronizing controls.

With the addition of what is known as a *chopping circuit,* two horizontal traces could be generated. The chopping circuit acts much like the square-wave calbrator circuit operating at an extremely fast rate. On positive excursions of the chopper, one signal input was presented and on negative excursions the second signal was displayed. Since the chopping rate was much faster than the sweeping rate and the signal frequency, two traces are generated each showing a different signal. This made it possible to analyze two waveforms together. Of course both traces had the same horizontal time base because only one sweep circuit is used. The philosophy of dual-trace generation to compare two signals by means of a chopper is shown in Fig. 1.22. Now phase measurements between two waveforms are possible by spreading one waveform between marks on the screen graticule and measuring the time difference in degrees of phase shift. Such a measurement is shown in Fig. 1.23.

1.4.1.4   The dual beam/trace oscilloscope.   To aid the analysis between two signals with differing frequencies, the dual-beam CRT was devel-

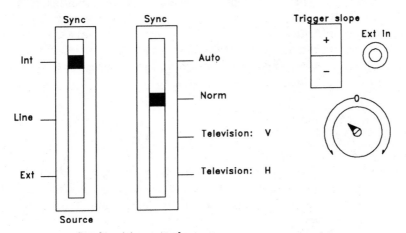

Figure 1.21   Synchronizing controls.

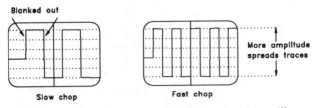

Figure 1.22 Development of dual trace. Faster chop will make two traces clearer.

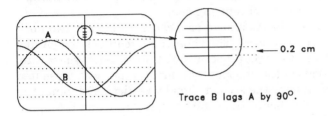

Figure 1.23 Measurement of phase.

oped. This tube was essentially two different tube elements sharing the same filament (heater) circuit and glass envelope. An oscilloscope using this type of tube had two separate vertical amplifiers and two separate time bases (sweep circuits). Therefore it was nearly two separate oscilloscopes sharing a common display showing both traces with two separate signals. Naturally this type of instrument was more expensive than the two trace instrument using a chopper.

Further development of CRT technology brought the variable persistance and storage scope. Even today some of these instruments are still in use. By varying the persistance a transient condition of the input waveform could be saved for analysis. It must be remembered that such instruments stored the waveforms in the electro optical domain. The arrival of digital storage, that is, solid state memories which can store the waveform information in digital form made obsolete the need for such expensive CRT displays.

1.4.2 Oscilloscope improvements

The modern-day oscilloscope is an extremely valuable piece of test instrumentation. It has been shown in the previous sections that essentially an oscilloscope takes a picture of the input signal, that is, shows an $x - y$ (two-dimensional) graph of this signal. Therefore signal distortion and noise can be estimated, while frequency, amplitude, and phase differences can be measured. Throughout the years more circuitry has been added to the basic oscilloscope to increase the accuracy and versatility of the instrument in measuring such things as pulse duration, duty cycle, pulse rise, and fall time, and over- or undershoot to name a few. Some of this circuitry will be discussed to illustrate this instrument's evolution.

1.4.2.1 Sweep triggering. Since the triggering circuits that start the sweep actually trigger on the input signal, the portion of the input signal before the triggering is not seen on the CRT display. By placing a delay line in the vertical amplifier signal path of the oscilloscope, the input signal will be delayed until the sweep gets going. Now the whole waveform is visible.

The triggering circuits of most of today's modern oscilloscopes have several controls. One is a button that selects the slope of the input signal that is desired to enable the trigger. The slope can be positive or negative. Another control selects the level of the signal when triggering is to take place. This control is a knob control marked − to + which can set the triggering point at a negative or positive signal polarity. Two-channel instruments have a selector switch to allow the sweep to be triggered on a selected channel. Also a single sweep can be selected. This enables the oscilloscope to be armed and waiting for an input signal to only trigger one sweep. This can be helpful in analyzing a transient signal condition.

1.4.2.2 The storage oscilloscope. To analyze the fleeting trace of a signal transient can be a frustrating experience when observing a basic oscilloscope screen. Some method of storing an oscilloscope trace was needed. First a variable-persistence type of oscilloscope was designed, with a special CRT. Earlier types of CRT memories used a bistable phospher coating on the screen. The two stable states were written and unwritten. Further development in CRT memory resulted in enhanced variable-persistence tubes. Such tubes had a storage mesh which stored the trace in the mesh behind the screen. Coupled with a bistable phospher screen and variable persistence, this type of storage oscilloscope is still in use today. Long storage time is still a problem with this type of screen and photographs of the screen are a means of permanent recording.

With the development of solid-state memories and integrated circuits it was not long before the arrival of the all-digital oscilloscope. The first step in the development of the digital oscilloscope was the introduction of digital controls. This essentially did not change the display or the method of driving the CRT. Push-button controls that controlled logic circuity that set the input attenuation and the sychronizing circuits used digital logic circuits. Using digital controls allowed the instrument to be set up through one of the standard digital interface busses by a digital memory or personal computer. Nat-

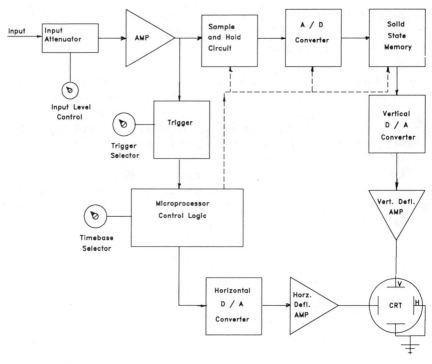

Figure 1.24 Basic digital oscilloscope.

urally the all-digital oscilloscope was just a step away. A block diagram of a basic digital oscilloscope is shown in Fig. 1.24. The brains of such an instrument are in the microprocessor control logic, which controls the sample-and-hold analog-to-digital conversion and the memory read/write, all according to the time base control setting. When the trigger circuit as selected by the trigger slope and level controls fires a start signal to the microprocessor, the process of sampling the input signal and analog-to-digital conversion commences. Much of the circuitry of such oscilloscopes is contained on only a few microchips, thus the cost of manufacturing has been reduced, resulting in lower prices to the user.

1.4.2.3 Oscilloscope accuracy. Accuracy and resolution are very important criteria in making a measurement. Accuracy simply refers to how close the measurement is to the true value, while precision refers to how much each measurement varies, that is, the repeatability of results. Resolution refers to how small a part of the input the instrument will see. Therefore, in order to obtain a higher degree of accuracy, one must be able to distinguish the small differences. In the

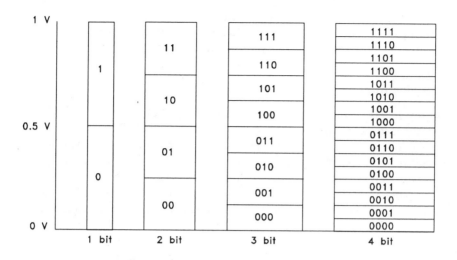

Resolution of 1, 2, 3, and 4 bit A/D conversion

Figure 1.25 A-D conversion.

digital domain resolution is directly related to the number of bits in the A-D conversion. The chart in Fig. 1.25 illustrates the point. Note that with each added bit, the number of levels the signal is divided has doubled. Therefore for each bit number, the number of levels goes in a 2,4,8,16 sequence. For an 8-bit word, the voltage is divided into 256 different levels, which corresponds to $\frac{1}{256}$ or 0.39 percent or $0.0039 \times 10^6 = 3900$ parts per million (ppm). These same rules apply to a digital voltmeter.

The time base in a digital oscilloscope is generated by a precision stable crystal oscillator, while in the analog type of instrument a ramp generator was used. The ramp generator's time base was set by the time constants of resistors and capacitors, which although temperature-compensated, were not as accurate as the crystal oscillator. Therefore most digital oscilloscopes have time base accuracies on the order of 0.01 percent, while analog instruments have a 1 to 8 percent time base accuracy. This type of time base accuracy is a big plus and is featured in TEK 2000 series oscilloscopes. The integrated circuits and CCD (charge-coupled device) memory are extremely fast and can acquire signal data at rates over 100 megasamples per second.

When either purchasing an oscilloscope or selecting a leased or rental unit, a good knowledge of the characteristics of the signal it is desired to measure is very important. For the most part a good general-purpose digital oscilloscope with a signal bandpass appropriate for the signal to be measured will do the job. However, in some instances for special transient types of signals, the manufacturers field application engineer's suggestions and advice can be extremely helpful in selecting the proper instrument.

1.5   Instruments with CRT displays

Many special types of instruments that use a CRT display are actually specialized forms of the oscilloscope. Since the CRT is basically an $x - y$ (horizontal axis versus vertical axis) type of instrument, it lends itself to a variety of applications. The oscilloscope for general purposes presents the display of signal amplitude versus time, and most oscilloscopes allow use of the vertical and horizontal amplifiers for a simple $x - y$ display. Amplitude modulation of a radio carrier can be measured using the $x - y$ mode of an oscilloscope. Also phase and frequency through the well-known Lissajous patterns uses the $x - y$ capabilities of the oscilloscope. The references at the end of this chapter explain some of the $x - y$ uses of the oscilloscope.

1.5.1   The television waveform monitor

A specialized type of oscilloscope used in the television industry is the waveform monitor. The time base of this instrument has fixed selections based on the video frame and field times. The horizontal synchronizing is rated at 1 and 2 times the horizontal rate.

More advanced types are able to select the horizontal lines in the vertical interval which allows the vertical interval test signals (VITS) to be analyzed. By analyzing these test signals, forms of picture impairment can be traced along the transmission path from the studio to the television transmitter.

1.5.1.1   The video waveform.   The video waveform is a complicated signal and problems with video frequency response can distort the horizontal synchronizing pulses and change the level of the chroma burst signal. Low-frequency-response problems can cause brightness changes and poor color saturation. To aid in analyzing the video signal response of a television signal, the VITS signals were developed. Two types of signals appear on the vertical interval, usually on line 17 or 18 of the lines appearing at the top of the television picture. These are the composite test signal and the combination test signal, as shown in Fig. 1.26a and b.

1.5.1.2   The video waveform monitor graticule.   The graticule on the screen of a waveform monitor is also more complicated than an oscilloscope graticule. The vertical scale is in what is known as IRE units,

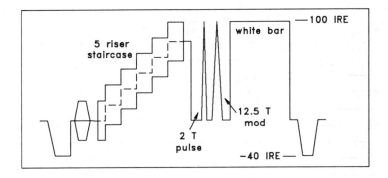

FCC Composite

(a)

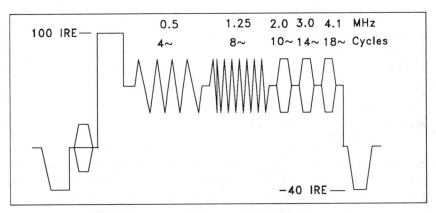

FCC Multiburst Combination

(b)

Figure 1.26   Video test signals.

with a maximum amplitude of 140 IRE units, which corresponds to 1-V peak to peak. This signal amplitude is spread from -40 IRE (horizontal sync tip) to $+100$ IRE, which is the maximum white level. Using this scale and displaying the two test signals, a variety of tests can be accurately performed which identify many of the standard picture impairment problems. An example of one of the more standard waveform monitor graticules is shown in Fig. 1.27. More on video testing will be presented in Chap. 4.

1.5.1.3 The vector scope. The waveform monitor has difficulty in making phase measurements. In the TEK 1480 a precision phasing

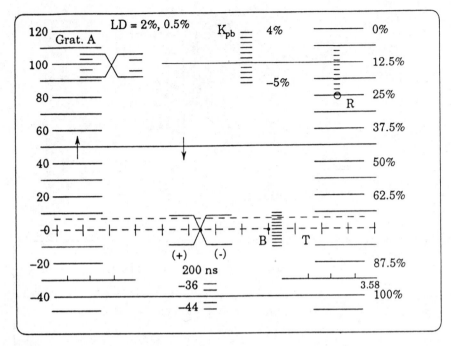

Figure 1.27 Waveform monitor graticule.

control section was added, which aided in making phase measurements, such as the differential phase parameter, much easier. However, in many simple monitoring applications, a simple vector scope will do the job. This essentially displays color bars which appear on one of the VITS signals as vectors with the center of the screen as an origin. The sweep is circular and is formed from 90° out-of-phase sine waves developed from the chroma subcarrier. The vector scope from many television broadcast facilities is a companion piece of instrumentation to the waveform monitor. The graticule for a vector scope is shown in Fig. 1.28. The standard connection for a waveform monitor and vector scope is shown in Fig. 1.29. It is extremely important that these instruments should be terminated in the standarded television impedance of 75 ohms. Essentially both of these instruments are high-impedance and have loop-through (input/output) ports so an external precision 75-ohm termination can be attached.

1.5.2 The spectrum analyzer

Another popular instrument using a CRT-type display is the spectrum analyzer. This instrument presents a display of amplitude versus frequency.

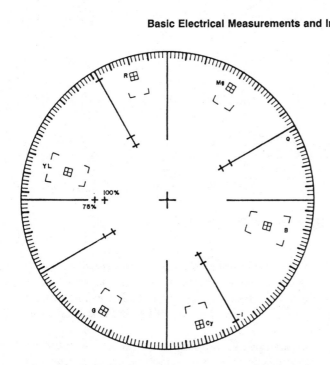

Figure 1.28   Vector scope graticale.

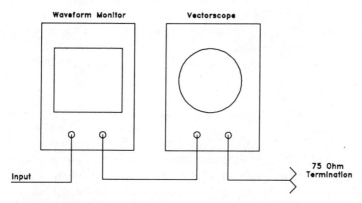

Figure 1.29   Typical waveform monitor-vectorscope connections.

1.5.2.1 Spectrum analyzer operation. Basically the spectrum analyzer is a signal receiver complete with detector which is swept
through its tuning range by the same sweep signal that controls the
horizontal oscilloscope trace. A signal that is connected to the input
test port will be seen on the scope trace when the sweep signal tunes
the receiver through the input frequency. The receiver circuits are often double or triple conversion with a selectable IF bandwidth. This
allows the area or span of frequencies per horizontal graticule division

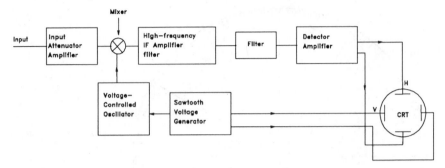

Figure 1.30   Block diagram of elementary spectrum analyzer.

to be selected so that the modulation side-band information can be seen and analyzed. Various types of filtering for the detected signal which increase the ability to identify signals out of the noise can be selected on most analyzers. A block diagram of the basic spectrum analyzer is shown in Fig. 1.30.

Not long ago most of the spectrum analyzers were either manufactured by Tektronix Corporation or Hewlett Packard Inc. However, several other manufacturers make more specialized equipment for the radio communications industry and the cable television industry. The Hewlett Packard model 8558 and the Tektronix 7L12 spectrum analyzer module plugged into a 7000 series oscilloscope mainframe were the so-called standard of the industry. The storage/memory system of the Tektronix 7613 oscilloscope enhanced the analyzing capabilities of the spectrum analyzer.

Presently Hewlett Packard makes the 8590 series instruments, which can be programmed via a magnetic memory card. Thus the instrument can have its personality program for communications, cable television, or EMI/RFI (electromagnetic interference/radiofrequency interference) work. Also the built-in memory can store data and can drive an ink-jet plotter for permanent data records. The Tektronix series 490 analyzers are extremely sophisticated instruments with a variety of options. Tektronix latest spectrum analyzer is the 2710 model, which has a true analog display as well as all the digital features, plus a variety of options.

1.5.2.2   Spectrum analyzer measurements.   Modern spectrum analyzers have the ability to spread out the frequency area of interest so that beat carriers that are the result of system distortion can be identified and measured. By examining the position and amplitude of side-band information, the condition of modulation can be analyzed. By placing an antenna array to a preamplifier bandpass filter combination, sources of EMI/RFI can be tracked down. Also, carrier-to-noise measurements can be made on many types of communications links such as point-to-point microwave, television broadcast reception, cable television systems, and computer local area networks (LAN) to name a few.

1.5.3   The time domain reflectometer

A useful instrument for finding open- and short-circuit conditions as well as impedance variations as a function of distance that often uses a CRT display is the time domain relfectometer (TDR). Such instruments are often specialized as to use on various cables. Tektronix model 1502/1503 uses a CRT display to test metallic cables such as electrical power cables, telephone cables, and coaxial cables.

1.5.3.1   Basic operation.   The principle of operation of a TDR is similar to a radar system, where a pulse of electrical energy is sent into one end of a cable. When this pulse of energy arrives at a short circuit or open circuit, a dent or break (a pulse of energy) is reflected back toward the sending end and received by the instrument which displays the result on a CRT. The polarity of the reflected signal indicates an open- or short-circuit condition. Figure 1.31 illustrates the screen presentation.

1.5.3.2   Measurements.   The operator controls that most present-day TDRs have available are the usual CRT controls such as brightness and focus plus a distance selector, a counter control and cursor, and a cable velocity factor or propagation constant selector.

The distance control is marked in increments of maximum range and selects the transmitted pulse amplitude and width. The wider and greater the pulse, the more energy is needed to search long cable lengths. Usually if the cable to be tested is long one should start with the long range. If a cable problem is seen closer to the instrument, a

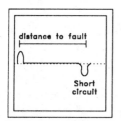
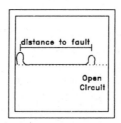

Figure 1.31   Reflected pulse shape on CRT for short and open circuit condition.

shorter range can be selected. Some TDR instruments have a chart recorder option which gives a hard copy of the measured results and avoids recording the screen trace by camera.

1.5.4 The network analyzer

The network analyzer manufactured by several companies uses a CRT display as an instrument readout. The network analyzer actually comprises several sections or separate instruments in a package, the oscilloscope being one of them. The oscilloscope of a microwave network analyzer such as the HP 8410 series* gives a polar display of amplitude and phase in polar coordinates. The signal generator portion is often a sweep-type generator which sweeps the signal through the frequency range of interest.

1.6 The LCD Display

Another type of instrument display quite popular in many instruments is the liquid crystal display (LCD) screen. This type of display is an outgrowth of the computer industry, which developed this display as a low-cost, low-power computer monitor. Many of the so-called lap-top computers have variations of the LCD display. Typically the display is not as fast-writing as a CRT. However, for lower-frequency oscilloscopes, TDRs, and various multimeters, this display type is becoming more and more used.

1.6.1 LCD screens and bar graph displays

The LCD display is often used to give a histogram (bar graph) display or dot/line display, as well as alphanumerics and movable cursors. Tektronx models 1502c and 1503c use an LCD display with trace storage capability which displays the transmitted pulse point and the reflected pulse. When the cursor is moved out from the transmitted pulse to the reflected pulse, the cable impedance and distance are printed on the LCD display. Front panel control scale factors also appear on the screen. A chart recorder option provides a permanent record.

Another instrument that uses a wide-screen LCD display is a signal-level meter manufactured by Comsonics Corporation called the Window.†

This LCD display gives a bar graph of the video carrier level and the audio carrier level with the values printed in dBmV on the screen.

*Trademark Hewlett Packard Co.
†Trademark Comsonics Corp.

Also the whole spectrum of television carriers, both video and audio, can be presented as a histogram display (multi bar graph) with cursors positioned to make specific measurements. The LCD display has caused many instrument types to appear on the low-cost instrument market, thus making many quite sophisticated instruments affordable.

2

Measurement of Electrical Quantities and Components

2.1 Electrical Parameters

Most measurements made on electronic equipment are the parameters of voltage, current, and power. These measurements are usually made during the final product testing phase or incoming testing before placing the equipment in day-to-day use and of course during system diagnostic and maintenance stages. Verification of proper operating voltages, current, and power indicate that at least the system has an operational power supply and that power regulation circuits are operating correctly. Oftentimes the main problem with many electronic systems is a poor or defective power supply and by monitoring the power supply's electrical parameters, a power supply failure and system outage can be avoided.

2.1.1 Voltage

One of the easiest measurements to make on a system is voltage. Voltage can be measured using a variety of methods, and a review of some of the references at the end of this chapter will tell of some of the earlier instruments and voltmeters. Instruments used in today's electronic industry consist of various types of voltmeters with accuracies from 5 percent of full scale to nearly 0.002 percent of reading. Many multimeters used today still use the electromechanical (D'Arsonval) meter movement as an indicating device. Many people still feel that a meter showing small flickering changes is easier to read than trying to read flashing digits on a digital meter. The digital voltmeter has to be powered with a power supply or batteries, while the D'Arsonval-type meter only needs a battery in the ohms mode.

It must also be remembered that any voltmeter is a high-impedance (resistive) device and that an ideal voltmeter draws no current from the circuit being tested. Modern digital voltmeters have input resistances that range from 10 megohms to nearly 10,000 megohms. Also in the ac voltage mode, low-input shunting capacity is necessary for high-frequency operation. The voltmeter is always connected across a circuit or component and one terminal of the voltmeter is always marked either-or common and is usually colored black with a black connecting wire and probe. The other terminal is usually marked + and has a red wire and probe. Some high-frequeny ac voltmeters have special hot or + marks on the cables and probes.

2.1.1.1 DC voltage measurements. Solid-state devices are always powered by direct currents and voltages. Also, vacuum tubes require dc voltage sources as well. Thus many measurements are made in the dc mode. Many systems today have built-in diagnostic meters so that power supply parameters can be monitored and logged. Figure 2.1 shows two common test connections involving dc voltage measurements.

Accuracies of a typical dc bench-type digital multimeter can be as good as 0.002 percent ±1 count of reading for a 6½ digit instrument. This accuracy is often available on the lower voltage ranges usually below 300 V; for higher voltages accuracies may be slightly worse.

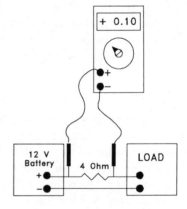

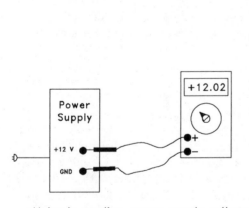

Meter is reading power supply voltage

Meter is reading the voltage drop across the 4 Ohm resistor. Current through the resistor is

$$I = \frac{0.1}{4} = 0.025 \text{ A} = 25 \text{ mA}$$

Figure 2.1 Voltage measurement

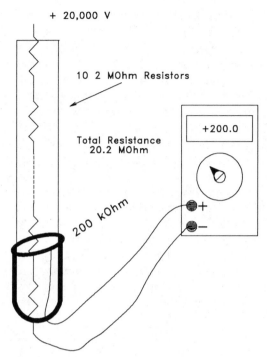

+ 20,000 V

10 2 MOhm Resistors

+200.0

Total Resistance
20.2 MOhm

200 kOhm

+

—

Figure 2.2 High-voltage probe, 100:1.

The point here is to read the instrument's specifications carefully, and if things are not clearly spelled out, contact the instrument manufacturers' applications engineering department.

High-voltage dc measurements usually involve a high-resistance voltage divider, which usually forms a high resistance to the circuit being measured and thus draws infintesimal current. One common need for measuring dc voltages in the range of 2 to 30 kilovolts (kV) is in measuring cathode-ray tube (CRT) final anode voltages. Many CRTs used in color television sets and computer monitors have voltages in the range of 15 to 30 kV. The accuracy of such measurements usually is not critical and should be in the range of 3 to 5 percent of the reading. A highly insulated hand-held probe containing the voltage divider is often used and comes as an accessory from some manufacturers. Such a probe is shown in Fig. 2.2 and is a 100:1 voltage divider. Most any high-impedance dc voltmeter can be used with such a probe and the voltage measured at the probe tip is 100 times that read on the meter.

2.1.1.2 Alternating voltage measurements. Alternating current (ac) voltage measurements require a more complicated voltmeter due to the many waveshapes the voltage may take. The most common is of course the sine wave. Commercial utility–supplied power is a sine wave with low harmonic content. Generally ac power generated by rotating machinery has the shape of a sine wave. Multimeters which use an electromechanical D'Arsonval meter movement have the ac voltage scale calibrated for a sine wave at 60 Hz. The frequency range is very small due to the distributed capacitance and inductance of the instrument circuitry and usually is good to possibly 100 to 150 Hz. Multimeters of this type have a rectifier circuit which essentially changes the ac voltage to dc, which is then read by the dc meter circuit.

Digital-type multimeters of present-day variety usually perform the ac measurement by digital computation. A single integrated circuit (IC) chip contains just about the whole digital multimeter circuitry. AC voltages are calibrated for sine wave measurements with frequencies up to about 1000 Hz. Many digital multimeters have the capability of reading ac voltage as true rms values. These instruments use circuitry that converts the ac voltage to the digital values by means of an analog-to-digital converter (ADC) and then computes the true rms value by a digital algorithm that integrates the area under the squared-value curve, over the period of one cycle followed by taking the square root. AC values of voltage at true rms may be measured by these instruments in the frequency range of up to 300 kHz for a sine wave. As a rule of thumb, for nonsinusoidal waveforms the bandwidth of the instrument should be 10 times the nonsinusoidal frequency. Therefore for a bandwidth of 300 kHz, nonsinusoidal waveforms of 30 kHz can be measured correctly at true rms.

Still in use today are thermocouple instruments capable of a frequency measuring ac values at true rms of up to 50 MHz. The method of measurement uses a thermocouple which essentially measures the heating value of the ac voltage, which in turn causes the thermocouple junction to produce a voltage measured by a D'Arsonval electromechanical meter or a digital indicating meter. This type of instrument has a usual accuracy of 1 percent and reads both dc and ac values.

For frequencies above 50 MHz and on up to 1 GHz (1×10^9 Hz) a peak-reading electronic instrument is used. Essentially this instrument uses a peak detector, the ouput is a dc voltage corresponding to the peak value of the very high frequency (VHF). The peak detector's storage capacitor is read by a very high impedance electronic voltmeter. Meters of this type can have high sensitivity for reading small rf voltages of very high frequencies to microwave frequencies. These meters have their scales calibrated for rms values for a sine wave and many also have a peak scale as well. The meter connections of this type of instrument are usually coaxial cables connecting the probe to the instrument. The probe may house a coupling capacitor or in some cases the detector diode which then supplies the dc value corresponding to the peak value to the instrument storage capacitor amplifier combination.

In many nonsinusoidal and complicated ac waveforms, it is often more desirable to merely measure the peak-to-peak value. A prime example is the NTSC (National Television Study Committee) base-band television video signal. This is the signal generated by a television broadcast camera which ultimately modulates the transmitted rf carrier. This waveform and its measurements are set at a standard 1-V peak-to-peak signal level which corresponds to peak white and horizontal synchronizing pulses. This 1-V peak-to-peak is specified as the peak-to-peak voltage across 75 ohms. Measuring nonsinusoidal waveforms for peak-to-peak values is usually made using an oscilloscope. The oscilloscope of the present-day variety can make peak-to-peak measurements on the order of 1 percent accuracy using the oscilloscope built-in voltage calibrator and precision attenuator/amplifier in the vertical channel. For the NTSC video waveforms, the peak-to-peak signal is measured in IRE units where 140 IRE units corresponds to 1 V peak to peak. Therefore 100 IRE units correspond to 0.714 V which is white level to blanking level. Measurements made on a special-purpose oscilloscope called a video waveform monitor, calibrated in IRE units can analyze the video signal for abnormalities that can affect picture quality.

2.1.2   Measurement of current

Current flow through a circuit or system allows the system to extract power needed to operate properly. Since current flow is through a circuit, an instrument measuring current flow should not have any resistance or impedance that will affect this current flow. Therefore current meters, that is, ammeters, should have minimal resistance, which is just opposite that of the voltmeter. The typical connections for current testing are shown in Fig. 2.3. A perfect ammeter should have zero resistance which of course is unobtainable. However, most ammeters used in practice are usually very low resistance.

2.1.2.1   The ammeter.   Most multimeters have both dc and ac ammeter ranges and for the most part have the ability to make common current measurements with sufficient accuracy. The meter leads connecting the meter to the circuit are the usual black/negative, red/positive, and, for example, the positive red wire should be attached to the most positive terminal and the negative to the most negative terminal so that

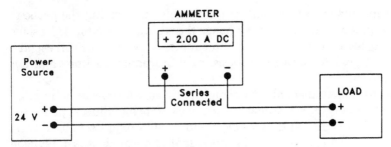

Figure 2.3 Current measurement.

the meter will be read as upscale. If connected with the wrong polarity, some digital ammeters will only indicate the polarity of the measured current.

A good convention to follow is to start with the highest current range and work down if the amount of measured current is unknown. Also, with a voltmeter, if the voltage level is unknown, one should start with the highest voltage range and work down.

To measure high values of current in the range of 10 A or greater, special high-range ammeters for dc are available. Some instruments used to measure current in hundreds of amperes involve using a low-value high-precision shunting resistor which is placed in series with the circuit load. The voltage drop across this resistor is measured with a sensitive high-resistance voltmeter. An example of this technique is shown in Fig. 2.4. The 50-mV shunts are usually available in sizes of 100 to 1000 A.

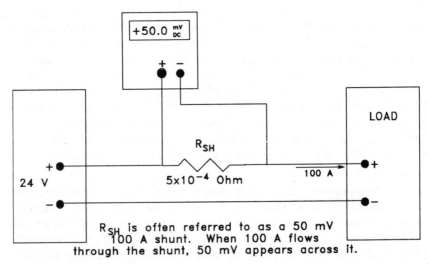

Figure 2.4 External shunt used for high current measurements.

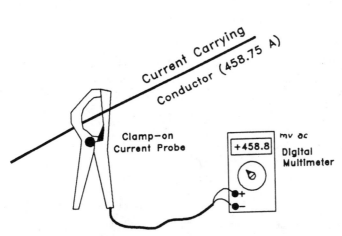

Figure 2.5 Clamp-on current probe measurement of dc current.

For dc current measurements that do not require high accuracy, a clamp-on dc current probe can be used with most multimeters that can measure in the range of a few millivolts. Such clamp-on-type devices do not require the circuit to be opened for insertion of the ammeter. The jaws of the probe are opened to place the current-carrying conductor. Often the output of the probe is 1 mV/A. Therefore, 10 mV corresponds to 10 A. The usual accuracy is on the order of 5 percent. Some more precise types can give accuracies of ±2 percent of reading. A test setup using such a probe is shown in Fig. 2.5. Current probes may have current ranges of 10 to 500 A.

2.1.2.2 AC ammeter. Alternating current is measured using essentially the same techniques as for direct current. However, instead of using a resistive shunt, a current transformer is used. A current transformer has a low-impedance winding which conducts the current to be measured. The secondary voltage is proportional to the primary current, which is in turn measured by an ac voltmeter of either the D'Arsonval or digital type. The ac clamp-on ammeter is actually a form of current transformer with an integral voltmeter in a self-contained unit. Accuracies for ac clamp-on ammeters range from about ±1 percent to ±3 percent of full-scale reading for electromechanical meters and 1 to 2 percent of reading ±1 count for digital clamp-on ammeters. Clamp-on ac ammeters usually can measure a frequency range of a sine wave of 40 to 400 Hz with the above accuracies.

For lower ac currents in the milliampere range, a current probe used in conjunction with an oscilloscope can be extremely useful in making current and phase measurements through a frequency range

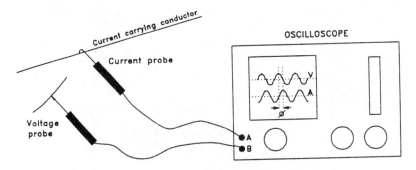

$\varnothing$ Is the phase difference between voltage and current

Figure 2.6   Oscilloscope current probe setup showing phase difference.

of 500 Hz to 50 MHz. This probe type has metal jaws that can be opened like the jaws of a clamp-on ammeter on the end of the probe handle. A push release opens the jaws to surround the current-carrying conductor. Such a probe type is shown in Fig. 2.6.

2.1.2.3 AC power measurement, the wattmeter. Measurement of ac power is usually done using many of the commercially made wattmeters. At normal powerline frequencies of 40 to 400 Hz, clamp-on current probes, along with the voltage probes used with the accompanying wattmeter, can measure ac power with accuracy of ±2 percent at unity power factor. Voltage ranges are normally 90 to 500 V and current ranges of 1 to 500 A result in measurements of power at about 200 kW. Instruments of this type are available for 3-phase, 4-wire power systems, and many give readings in kilowatthours with the option of adding a chart recorder to monitor power usage. Using both current and voltage probes, a hand-held power factor meter is commercially available with accuracies of ±0.2 to 1.5 percent for both leading and lagging power factors. Such instruments are very useful in searching out electrical equipment responsible for poor power factor operation, which could result in excessive electrical power charges. The connections for a 3-phase power meter are shown in Fig. 2.7.

An instrument used in radio and television receivers as well as many types of consumer electronic equipment is a combination of an adjustable isolation transformer, ammeter, voltmeter, and wattmeter manufactured by Sencore Corp. This instrument is referred to as the PR-57 Powerite. The isolation transformer is connected to the commercial power and its secondary feeds power to the instrument under test, thus isolating the commercial power ground from the chassis ground of the instrument being tested. This allows for sensitive instruments to make measurements on the device without the possibil-

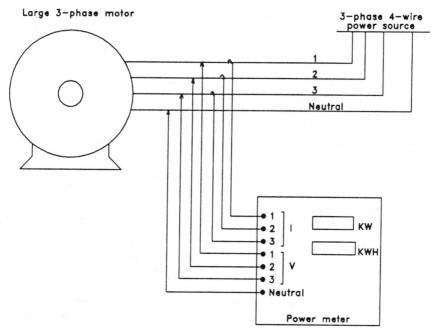

Figure 2.7 Three-phase power measurement using power meter with current and voltage probes.

ity of having the instrument case connected to one side of the power line. Metering of the voltage or current and power feeding the instrument under test can indicate an improperly working power supply or power regulator circuit. This instrument can also measure any high-resistance leakage path between the chassis ground and the power system ground, which can forewarn of any breakdown that could cause a shock hazard. A properly working piece of equipment should draw the correct amount of power as specified by the manufacturer when the correct voltage is supplied. Also, negligible leakage between the chassis ground and the power neutral should be measured. By measuring the voltage, current, and power, the power factor can be calculated by the usual formula pf = watts/volts × amps. Usually for electronic equipment that uses a transformer at the input of its power supply, a lagging current will be drawn and hence a lagging power factor results. Normally power factors are really quite high (better than 70%) for most transformer operated equipment.

2.2 Resistance, Capacitance, Inductance

The measurement of the circuit elements of resistance, capacitance, and inductance is as important as the measurement of the electrical quantities. It is expected that accurate circuit elements can be purchased from manufacturers, and when discrepancies are found, often the method of measurement may be the reason. When resistors are manufactured, one would expect that the resistance is pure over a wide frequency range. The connecting lead inductance and the distributed capacity have a decided effect at high frequencies. Therefore, to be accurate, measurements at various frequencies should be made to test for the effects of inductance and capacitance. Also, capacitors have lead inductance and some high-resistance leakage, which can have an effect on the circuit operation at high frequencies. Capacitors are available in a multitude of types such as Mylar, mica, electrolytic, tantalum, and ceramic, to name a few. Whether they are polarized or nonpolarized, no one method of testing will tell the whole story. Large values of capacity usually are electrolytic of aluminum or tantalum variety and are polarized with positive (+) and negative (−) terminals marked. If the polarity is not observed, the capacitor can be destroyed with catastrophic results such as overheating and exploding or leaking electrolyte over the circuit components. At high frequencies inductances have distributed capacity and resistance of the turns of wire that make up the induction coil. Manufacturers often test particularly inductors and capacitors on an rf bridge instrument which tests for the higher-frequency effects and specifies such parameters as the Q factor and dissipation factor as well as inductance and capacitance.

2.2.1 Resistance measurement

The measurement of resistance can be made by a variety of instruments depending on whether the resistance range is extremely low, high, or requires at appropriate accuracies high-power, or at high-frequency or just the dc value.

2.2.1.1 The ohmmeter. For most applications of normal circuit work the ohmmeter function of most digital multimeters is sufficient. Since resistors of the carbon, carbon film, or metal film variety used in most electronic circuitry have accuracies ranging from 1 to 10 percent and many digital multimeters have resistance measurement accuracies of ± 0.25 to ± 0.5 percent, the resistance values can be properly confirmed for most applications. When measuring resistances with an ohmmeter, care must be taken to make good electrical connections to the resistor leads to assure accurate measurements. Ohmmeters use a battery which causes a current to flow through the resistor to be measured. Therefore the measurement of the resistance is made at direct current. Most high-power resistors are often wire-wound and are used in dc or low-frequency ac circuits; therefore the normal ohmmeter is sufficient. Variable resistors can also be measured with an ohmmeter

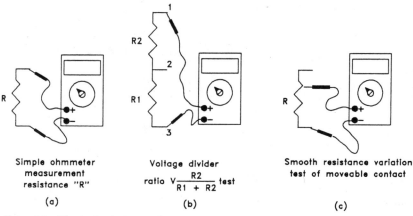

Figure 2.8 Ohmmeter test procedures

as far as the maximum value is concerned, also the amount of variation can be confirmed as well as the quality of the variable contact.
Often noisy contacts or an intermittent contact can be found using a
simple ohmmeter test. Some of the more common ohmmeter tests are
shown in Fig. 2.8.

2.2.1.2 The Wheatstone bridge. For more accurate testing of resistors
at direct current, a Wheatstone bridge can be employed. Accuracies of
about 0.1 percent are attainable with many of the instruments available. A schematic diagram of a Wheatstone bridge is shown in Fig.
2.9. The ratio of R_1 to R_2 is usually controlled by a decade switch
which may select the ratio 10^{-3}, 10^{-2}, 10^{-1}, 1, 10^1, 10^2, etc. This ratio
is selected and R_3 is varied, usually by a continuous control, to achieve
a balance. The value of the unknown resistance is read off the instrument's control dials. Operation of a bridge usually is a bit time consuming and a good study of the instruction manual often can speed up
operation and improve accuracy as well. Wheatstone bridges commercially available can make resistance measurements that range from 1
ohm to very high resistances of several megohms.

2.2.1.3 Measuring low resistances. Measurement of very low resistances can present some different problems. The quantity called *conductance* is the reciprocal of resistance and is given the symbol G.
$G = 1/R$, and the SI unit of conductance is the siemans (S). It should
be obvious that high conductance (good conductivity) corresponds to
low resistance, so if $G = 10$ S, then $R = 1/10$ ohm = 0.1 ohm. Some
sensitive digital multimeters are able to make measurements in
siemans in the range 10 to 10^{-8}. For the measurement of resistance
values from 10^{-5} to 1 ohm, a Kelvin bridge can do the job. The Kelvin

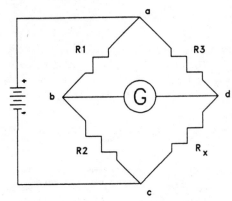

At the balance point, b and d
are at the same potential.

Resistors R1, R2, and R3 are known,

$$\frac{R_x}{R3} = \frac{R1}{R2} .$$

R_x is unknown resistance

G is the galvanometer

Figure 2.9   The Wheatstone bridge schematic diagram.

bridge is actually a modified Wheatstone bridge. For measurement of low resistances, the connecting leads and terminal resistances can greatly affect the measurement accuracy. The Kelvin bridge was improved by adding a second set of ratio arms, and this version of the instrument is called the Kelvin double bridge. The circuit configuration is shown in Fig. 2.10. This is a laboratory piece of equipment and should be kept clean, covered, and at an even temperature. Again, any use of a sensitive bridge instrument should include a careful study of the operations manual. This type of instrument is commercially available, but not many manufacturers produce the instrument.

2.2.1.4 Measuring high resistances. Resistance measurements in the very high resistance values require higher testing voltages. Resistances on the order of 100 to 2000 megohms can be measured with accuracies of 1.5 to 10 percent of reading. Resistance testers that measure resistances this high are often referred to as insulation testers. The test voltages often range from 250 V dc to 1000 V dc and the

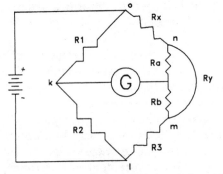

Ry is the yoke resistance.
Ra and Rb are the second set of bridge
arms which eliminate the effect of
the yoke resistance.

$$\frac{Ra}{Rb} = \frac{R1}{R2}$$

Figure 2.10   Kelvin double bridge configuration.

power sources may be a hand-crank high-voltage generator or a battery-operated internal dc-to-dc converter. Caution should be exercised when using such equipment because a shock hazard does indeed exist. When using such equipment to measure very high resistances, the operating manual should be thoroughly studied.

2.2.2 Capacitors

Since capacitors come in so many sizes and types, no one perfect method for measurement is practical. A good discussion of capacitor types is given in several of the references listed at the end of this chapter. As the frequency of the applied ac voltage to a capacitor increases the effects of lead inductance come into play. The high-frequency model of a capacitor is given in Fig. 2.11 where L is the lead inductance, R_s is any lead and connection or series resistance, C is the capacitance, and R_c is the dielectric leakage of the capacitor. If R_c is infinite, then the capacitance C would hold a charge indefinitely, which is not possible. The method chosen to test a capacitor should, for all practical purposes, be close to the manner the capacitor will be used in a circuit. Since some capacitors are polarized, the terminals are marked + and –. Therefore any alternating voltage applied should have a dc component that disallows any polarity violation. Polarized capacitors usually come in larger capacitance values from one to several thousand microfarads with applied dc voltages of 1 to 1000 V. As the value of capacity increases, usually the dc voltage rating across the capacitor decreases. When testing such capacitors, any test voltage should be less than the dc voltage rating of the capacitor.

2.2.2.1 Capacitance meter.
One of the currently used and low-cost methods for capacitor testing is with one of the new capacitor testers

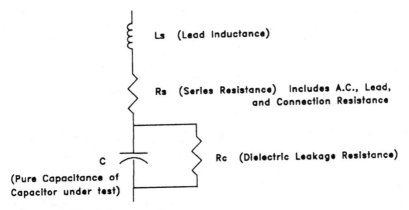

Figure 2.11 Circuit for a capacitor.

manufactured by several companies. These instruments can measure capacitors in the range of about 1 pF to nearly 200,000 μF with accuracies ranging from 0.5 to 3 percent of reading. Since some instruments are more accurate than others and the more accurate ones are more expensive, it pays to examine one's needs closely before deciding which type to purchase. The prices generally range from about one hundred dollars to several hundred. Many such instruments have the capability of also measuring inductance and resistance as well and are appropriately called LCR meters or impedance meters. Accuracies of measurement for resistance range from 0.5 to 3 percent and for inductances about 2 to 3 percent of the value. Most of the lower-priced instruments are battery-operated and are about the same size as some of the smaller digital multimeters and thus can be placed in a field engineer's or technician's tool kit. Many digital multimeters have a capacitance tester built in as an added feature. Digital techniques are used in these types of instruments, which essentially measure the charge or discharge time of the capacitor under test. The capacitor is allowed to charge up to an accurate reference voltage, and a gate pulse is started when the capacitor begins to discharge to a lower accurate voltage level through an accurate known resistance. When the capacitor discharges to a lower reference voltage, the gate signal is closed. The time interval of the gate voltage is a measure of the RC time constant of the unknown capacitor and the precision resistor. A clock frequency and counter circuit measure the time constant. Then the capacitance can be calculated and a numerical reading of capacitance is shown on a digital readout. If a resistor is substituted for the capacitor under test and a known accurate capacitor is interchanged for the internal resistor, then this type of instrument can be used to measure resistance. An instrument of this type usually is used to test occasional capacitors in a maintenance or repair program. For production-line testing, instruments are available that, when connected to a testing stage where the finished capacitors are allowed to pass through a microprocessor controller or computer system, can signal the capacitor tester to make a measurement. The capacitor tester presents the measurement in digital form to an IEEE or RS232 digital port, which signals the computer to record the measured value or simply reject or accept the capacitor. One type of production-line test for capacitors is shown in Fig. 2.12.

2.2.2.2 High-frequency capacity testing. At high frequencies the effects of lead inductance, series resistance, and leakage resistance become more pronounced. Therefore it is reasonable to test the capacitor by some high-frequency means. An ac bridge circuit can be operated at various frequencies, and measures the capacitor for capacitance as

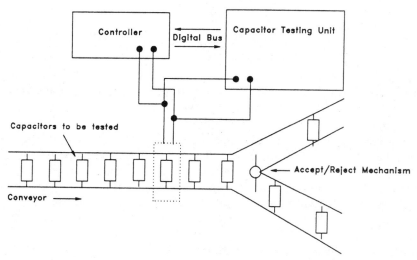

Figure 2.12 Automatic testing for capacitors using digital capacitor tester and digital controller.

well as the effects of leakage and series resistance and inductance. Essentially the configuration is the same as the Wheatstone bridge, except the battery is replaced by an ac source at the desired frequency and the meter circuit is replaced by a sensitive detector to detect the null that indicates the bridge balance. The bridge resistors are replaced by impedances which may consist of series or parallel combinations of inductances, capacitances, and resistances. Such configuration of impedances determines the name of the bridge circuit such as a series comparison or parallel comparison bridge or a Schering bridge. The conditions for a bridge balance must be met in order to balance the bridge circuit necessary for a measurement of the unknown element. Many technical school laboratories have such various types of bridges set up to teach bridge theory. However, commercially manufactured bridges are available and the type of bridge circuit employed depends on what type of element is to be measured, that is, capacitors or inductors. Commercial capacitance bridges can measure capacitor values from 1 pF to 1000 μF with accuracies ranging from 0.1 to 1 percent of the value. Many present-day commercial capacitance bridges employ digital readouts of capacitance and dissipation factor. It must be remembered that a bridge is basically a comparison method of measurement, where impedance ratios balance the bridge. Operation of a bridge involves a number of controls that are used to obtain the balance condition. This is time consuming unless the same value of capacitance is to be measured. Bridge circuits have been employed in production-line testing where the bridge is balanced using an acceptable value of capacitance. The limits of unbalance are used to either reject or accept the capacitor under test.

Larger-value capacitors are often of the polarized electrolytic type, usually aluminum or tantalum. The tolerance of such capacitors is usually large and may be marked, for example, as +80%/ − 20%. Therefore a 100-μF capacitor can range from 180 to 80 μF. Digital capacitor testers that measure the capacitor value by measuring its charging or discharging time usually give a higher reading than the marked value. This value of capacitor is how the capacitor will perform in dc circuits. Oftentimes production-line testing of electrolytic capacitors uses an ac bridge, where the signal source is, for example, 1000 Hz; the value of measured capacitance is then used to mark the capacitor value. It has been found for aluminum and tantalum electrolytic capacitors that the value changes with frequency and the measured value is usually lower than the actual value. Therefore if this method is used to measure and mark newly manufactured capacitors, then the value as measured with a digital capacitance meter will be higher. Usually silver mica, Mylar, and ceramic capacitors, when tested with either the bridge or digital meters, will give the same results.

2.2.3 Inductances

Inductances can also be tested in much the same way as capacitors, since they are mirror images of each other. Since inductors are made from turns of wire of various composition and size, the series resistance can vary considerably. The test method for inductances should be chosen close to the frequencies of the circuit using the inductor.

2.2.3.1 Inductance meter. As stated before, many of the digital capacitor testers also have the ability to measure inductances. The range of inductance measurement for such instruments is 1 microhenry (μH) to about 200 H with accuracies ranging from 1 to 5 percent. One of the digital L-C meters, called the Z-meter,* is unique in that inductances are measured by a ringing current waveshape. The number of cycles an inductor resistor circuit produces after being pulsed is a measure of inductance. Any shorted turn in the inductor will show up with this instrument. This instrument also has the capability of testing such active devices as silicon controlled rectifiers (SCR) or Triacs.

2.2.3.2 Inductance bridges. Inductances can also be measured by many ac bridges. The Maxwell bridge is particularly useful in mea-

---

*Trade name, Sencore Corp

suring inductances for low-inductor Q values below 10. The Hay bridge is useful when measuring high-inductor Q factors such as 10 to 100. It must be remembered that the Q factor is a function of frequency and mathematically given as

$$Q = WL/R \quad \text{where } W = 2\pi f$$

For high Q, R should be low, which means that the inductor should be wound with low-resistance material and fewer turns, which means lower inductance. As frequency increases, so does Q, provided the ac resistance remains constant. The bridge configuration for the Maxwell and Hay bridges are shown in Fig. 2.13. Many commercial bridge circuits are manufactured specifically to measure either capacitance or inductance. Bridges that are set up to measure both capacitors and inductors are often referred to as impedance bridges. Lower-priced impedance bridges are manually operated, and the more elegant models are self-balancing with digital readout; some even have computer communication busses (RS-232 or IEEE) for control and recording of results. Commercial inductance bridges are usually able to measure inductances in the range of 1 μH to 1000 H with accuracies of about 1 percent. Most bridges also give a measurement of the Q factor as well. As stated before, $Q = WL/R$, where L is the inductance in henrys and $W = 2\pi f$, where f is the frequency in hertz. R is the effective ac series resistance of the inductor and is also frequency-dependent. Therefore a measurement of R at direct current with an ohmmeter, and the

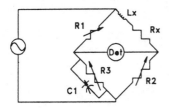

R1, R2, R3, and C1 adjusted for balance,

$Rx = \dfrac{R1 \; R2}{R3}$ and $Lx = R1 \; R2 \; C1$.

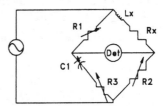

R1, R2, R3, and C1 adjusted for balance,

$$Rx = \frac{R1 \; R2 \; R3 \; \omega \; C1}{1 + (\omega \; C1 \; R3)^2}$$

$$Lx = \frac{R2 \; R3 \; C1}{1 + (\omega \; C1 \; R3)^2}$$

Also, $Q = \dfrac{\omega \, Lx}{Rx} = \dfrac{1}{\omega \; C1 \; Rx}$

At $Q = 100$, $\omega C1 \; R3 = .01$

so $Lx = R1 \; R2 \; C1$ (approximately)

(a)   (b)

Figure 2.13 (a) Maxwell bridge; (b) Hay bridge.

bridge measure of inductance, the calculated value of Q is quite different from the actual value. The best method is to make a measurement of Q directly. This is done with an instrument called a Q meter, although many commercial inductance bridges can also measure Q. At high Q values, commercial bridges' accuracy for the measurement of Q decreases.

2.2.3.3 The Q meter.   The Q meter is a fairly simple device, and unlike a bridge which has to be balanced with a single-frequency ac signal source, this instrument merely finds the resonant frequency of the inductor and an internal capacitor. The circuit for a Q meter is shown in Fig. 2.14, which includes the mathematical theory of operation. The meter reading the capacitor voltage has its scale calibrated directly in Q. The variable ac generator should have a low-impedance to drive current through the low-impedance circuit at resonance. Also, the voltmeter across C should be very high impedance compared to the capacitive reactance at resonance.

2.3 Impedance Measurements

The measurement of impedance is an extremely important parameter that should be measured often. All electrical terminals have an impedance appearing across the terminals. Devices with only one pair of terminals are known as two-terminal devices, and either constitutes an instrument's measuring terminals or a load or terminating impedance. Devices with two-terminal pairs, known as four-terminal de-

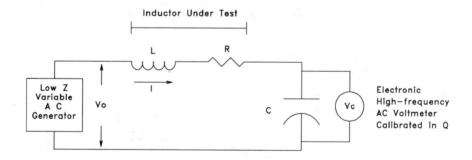

Vary f_o until voltage peaks across C indicate maximum current.
Now, $X_C = X_l$ and their effects cancel at resonance.

$$Q = \frac{V_z}{V_o}$$

V_o is fixed by the low-impedance source, therefore the meter scale can be calibrated in Q factor on Vc.

Figure 2.14 Diagram of a basic Q meter.

vices are usually active devices such as amplifiers or passive devices such as attenuators, or filters. The input terminals have an input impedance and the output pair has an output impedance. A piece of cable has or is supposed to have the same input and output impedance hence known as impedance matching which exhibits low reflections and maximum power transfer. The impedance Z has a real part and an imaginary part and is mathematically shown as

$$Z = R + jx \quad \text{in rectangular coordinates}$$

and

$$Z = \sqrt{R^2 + x^2} \quad \angle\theta$$

where $\theta = \pm\tan^{-1} X/R$.

Since reactance X, which is either inductive or capacitive, is a function of frequency, so then is the impedance. Therefore when making the measurement of either input, output or load impedance one must make the measurement at the specific frequency of interest or at all frequencies from 0Hz (dc) to whatever is the upper limit of either the instrument capability or the device.

2.3.1   Instrument input/output impedance

Most devices have published specifications for the input and/or output impedances. Seldom does a manufacturer ever say whether an actual measurement has been made or if so, what method was used to make the measurement. When coupling instruments and devices together to either make up a system or test a setup, any impedance mismatches will cause loading effects. These loading effects can cause errors in measurements.

2.3.1.1   Instrument circuit loading/oscilloscope probe.   The effect of loading and the errors it may produce are shown in Fig. 2.15. The signal source has a 10-V output and a 100-ohm internal impedance and it is desired to measure the 10-V value with various voltmeters with several values of resistance. It is evident that a high-resistance, that is, near-perfect, voltmeter does not load the circuit and cause an erroneous measurement. Most digital multimeters have significantly high internal impedances and hence minimize the loading effects. Oscilloscopes which basically measure signal voltages have input impedances typically of 1 megohm in parallel with about 25 pF of capacitance. With the additional capacitance of the cable, coupling the oscilloscope to the measurement probe often adds nearly 100 pF of shunting capacitance. To overcome these effects a compensating probe, which essentially adds a peaking network, offsets the shunting capacitive reactance.

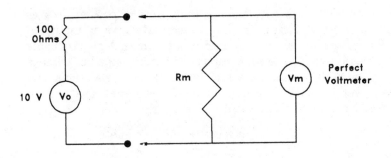

Rm = voltmeter resistance

Vm = measured voltage with a perfect voltmeter

Rm (Ohms)	Vm (Volts)	% Error
10	0.909	90.9
100	5.000	50.0
1000	9.090	9.1
100,000	9.9	0.1
Infinite	10.000	0.0

Figure 2.15 Meter circuit loading

Such a probe is often called a *low-capacity probe* and is shown in Fig. 2.16. Such probes have added resistance in series, which affects sensitivity but does decrease the loading effect. At point B in the figure the impedance of the cable and the oscilloscope is 1 megohm in parallel with C_s and C_c. When C_s = 50 pF and C_c= 15 pF/f, C_c for a 6-ft cable is 75 pF. The cable and oscilloscope's input looks like 125 pF in parallel with 1 megohm. Adding the series combination of the probe's resistance and capacity, where a 9 to 1 value is set for the probe's capacity, results in a total capacity of 12.5 pF. Recall that capacitors in series are calculated by product over sum. At the probe tip the circuit looks like a resistance of 10 megohms in parallel with 12.5 pF. As can be seen, the loading effects of the oscilloscope are reduced. In order to adjust the probe's compensating capacity, most oscilloscopes have a built-in square wave calibrator with a precision level and rise time. When the probe tip is touched to this test point, the probe capacity is adjusted with a small insulated screwdriver through a hole in the probe until the square wave signal has sharp vertical edges and no over- or undershoot. Also the precise level of this signal can be used to calibrate the vertical amplifier circuits. Oscilloscope probes usually have an instruction manual included and give use and adjustment information.

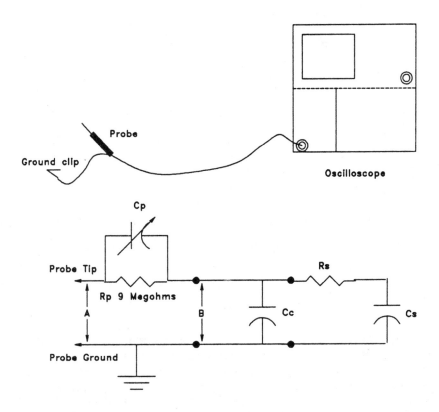

Cp = Probe Compensating Capacitance
Rp = Probe's added Resistance
Rs = Oscilloscope Resistance, typically 1 Megohm
Cs = Oscilloscope Shunting Capacitance, typically 25–50 pF
Cc = Cable Capacitance, typically 15 pf/foot or 50 pF/m

Figure 2.16   The oscilloscope low-capacitance probe.

2.3.1.2 Measurement of input/output impedance.

Most test instruments make known the test terminal impedance, since it is an important parameter for the whole test procedure. Many instruments display the input or output impedance on the front panel near the terminals. Many devices and systems to be tested usually have the terminal impedances published on their specification sheets.

If the impedances are questioned or it is desired to test them, a test method will then be selected depending on what test instruments are available. If measured by a simple digital multimeter, input impedances (resistances) will just be the dc value. If the device will only be used with direct current, then this test will be sufficient. Active de-

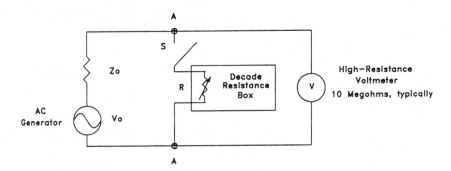

Zo is output impedance of generator.

With switch "s" open, read V for a measurement of Vo.
Close s and vary R until V reads half its previous value.
Now, R = Zo and can be read from the dial on the decade resistor.

Example: With s open, the meter reads 100 V. With s closed,
the meter reads 50 V. When R is adjusted to 1 kOhm,
R = Zo = 1 kOhm.

Figure 2.17 Source impedance test.

vices that provide an output signal, such as a signal generator or an amplifying device, use different procedures to measure the output impedances. Devices that provide a signal or electrical output can have their output impedance magnitude measured by the drop-in-potential method. This method is similar to the one used to find the internal resistance of a battery and is shown in Fig. 2.17. Although not very accurate, this method can still provide useful information about the terminal impedance. To expand on this method—if the device under test is a generator with adjustable frequency, then the output impedance can be measured for various frequencies points and a plot of output impedance versus frequency can be made. If the device under test has a high power output and a low internal resistance, such as a battery, then the current through the internal resistor may become prohibitive when the voltage across this resistor is halved. This problem and a solution to it are shown in Fig. 2.18. Adding the ammeter and only partially loading the circuit can do the job. An oscilloscope can be substituted for the voltmeter to record the drop in voltage. This gives the added benefit of making it possible to observe any waveform distortion during the loading precedure. When using an oscilloscope to monitor the voltage across the loading resistor, the 10:1 probe should be used to isolate the oscilloscope from the circuit.

2.3.2 High-frequency effects of instrument cables

As previously discussed, the loading effect on the input/output impedances can cause errors. At high frequencies, reflections caused by the

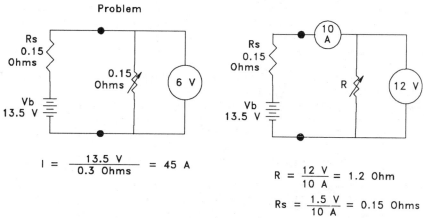

Figure 2.18 Measurement of battery internal resistance.

connecting cables and mismatched impedances can cause large errors. At frequencies above 10 MHz, cables and connectors to oscilloscopes, voltmeters, and spectrum analyzers are usually selected to be at 50 ohms impedance. The cable television industry, along with the television broadcast industry, has selected 75 ohms as the standard input/output impedance.

2.3.2.1 Impedance mismatch, reflections. The reflections caused by mismatches at high frequencies can cause large errors, since standing waves occur along the connecting cables or transmission lines. Such a problem is illustrated in Fig. 2.19, where the mismatch is ±2 ohms from the 75-ohm characteristic impedance of the cable. The standing waves, which are variations of rms voltage and current along the transmission line, give errors in measuring any load or generated signal. The case shown in Fig. 2.19 is not severe since the mismatch is so slight. If the generated signal is great enough and the reflections large enough to cause problems, addition of an attenuating pad isolating the mismatch can be placed in the line. The value of the attenuating pad will decrease the return loss by that amount. Where a great number of interconnecting cables are involved, care should be taken to avoid cables at multiples of a quarter wavelength at the test frequency. Any mismatches at the connectors will make matters worse.

2.3.2.2 Impedance match testing. Since mismatches at high frequencies cause reflections and reduce the power transfer from the generator to the load, the matching impedances should be tested. The usual tests involve an rf bridge, a sweep signal generator, and a detector and an oscilloscope which measures return loss. The impedance mismatch may then be calculated. A test setup used to measure return loss is

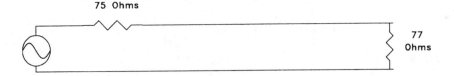

Lossless Transmission Line

75 Ohms

77 Ohms

The Reflection Coefficient is $K = \dfrac{Zr - Zo}{Zr + Zo}$.

$K = \dfrac{77 - 75}{77 + 75} = 0.013.$ 1.3% of the power is reflected.

To find the Voltage Standing Wave Ratio (VSWR),

$K = \dfrac{VSWR - 1}{VSWR + 1}$, 0.013 VSWR + 0.013 = VSWR − 1
0.987 VSWR = 1.013
VSWR = 1.026 : 1

Return Loss (dB) = 20 log $\dfrac{1}{K}$ = 20 log 0.013 = −37.7 dB

The reflected signal is 37.7 dB less than the transmitted signal.

Figure 2.19   Effects of transmission line impedance mismatch.

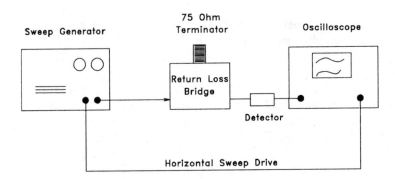

Example: If Return Loss is −25 dB,

$RL = 20 \log \dfrac{1}{K} = -25$

$\log \dfrac{1}{K} = 1.25$

$K = 0.056 = \dfrac{Z - Zo}{Z + Zo}$

$\dfrac{Z - 75}{Z + 75} = 0.056$

Z = 83.9 Ohms

Figure 2.20   Measuring return loss to determine device impedance.

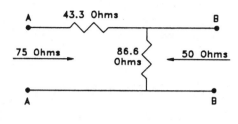

$$Z = \sqrt{Z_{oc} \, Z_{sc}}$$

For terminals A–A

Z_{oc} = 43.3 Ohms (B–B short Circuited)

Z_{sc} = 43.3 + 86.6 = 129.9 Ohms
(B–B open Circuited)

$Z = \sqrt{Z_{oc} \, Z_{sc}}$ = 75 Ohms

For terminals B–B

$Z_{sc} = \dfrac{(43.3)\,(86.6)}{43.3 + 86.6}$ = 28.85 Ohms

Z_{oc} = 86.6 Ohms

$Z = \sqrt{Z_{oc} \, Z_{sc}}$ = 50 Ohms

Figure 2.21 Terminal impedance by open- and short-circuit
method.

shown in Fig. 2.20. Also a Q meter, if one is available, can measure
the terminal impedance of a cable or device. At higher frequencies,
the stray capacities and inductances can make accurate measurements difficult. A good commercial bridge can measure the capacitance, inductance, and resistances of a terminal pair and the resulting
impedance calculated. Also some LCR meters can do essentially the
same thing.

For passive four-terminal networks such as attenuator pads, resistive
matching networks, and filters, the input and output terminals can be
measured with the opposite terminals properly terminated. For a pair of
terminals to be measured, if the terminating impedance of the opposite
pair of terminals is unknown, the following equation can be used. This
formula is often used in transmission lines where impedance of the terminals to be measured is the geometric mean of the terminal impedance
with the opposite terminals short-circuited and open-circuited. Figure
2.21 for the 75-ohm to 50-ohm impedance matching pad illustrates this
method. This method is used to measure the characteristic impedance of
transmission lines, pads, and filter networks.

3

Measurements
at Audio Frequencies

The word *audio* means "to hear" and has to do with audible sounds. Sound or audio waves produce complex electrical signals in electromechanical transducers such as microphones, loudspeakers, and phonograph cartridges. These transducers convert the mechanical energy to electrical energy and often are principal causes of audio distortion. The simple means of picking up sound waves by a microphone, amplifying the resulting electrical signals to a higher power level, and converting back to sound waves through a loudspeaker allows many things to go wrong, mainly because of the two conversions to the mechanical domain from the electrical domain. Measurement of audio signals in the electrical domain can point to defective electromechanical transducers as well as defects in the electronic amplifying systems. The radio and television industry make many ongoing measurements of their audio signals in the transmission chain of studio to transmitter. The major networks, through the use of telephone lines, microwave radio, satellites, and fiber-optical systems measure the audio signals along with their video signals. The telephone companies perform many audio quality measurements necessary to provide high-quality service to their customers. The consumer electronics industry, including radio and television service; FM stereo transmission; and disk, tape, and optical disk recorder service and repair all have to make measurements of audio level, distortion, noise and low-frequency disturbances with sufficient accuracy so that the customer does not notice or hear any audio impairments.

3.1 The Audio Signal

The audio signal for instrumental music as well as the accompanying sound from motion picture audio tracks is an extremely complicated waveform. Large changes in both frequency and level make faithful amplification and reproduction a formidable task. The study of Fourier analysis says that any periodic waveshape can be represented by a series of sine waves with various amplitudes and frequencies. In other words, an arbitrary complex waveform can be decomposed to a set of sinusoidal signals. It has been shown that a square wave can be represented by a series of sinusoids from the fundamental and the odd harmonics. Therefore testing of audio equipment using a square wave test signal is often performed, which gives a measure of the device's transient response. Also use of sinusoidal signals swept over the audio band of 20 to 20,000 Hz can provide a frequency response measurement.

3.1.1 Audio system standards

One of the standard line impedances for microphone lines and signal transportation input lines in most broadcast studios and recording studios is 600 ohms balanced. High-impedance microphones often have built-in matching transformers to match to the standard 600-ohm balanced line. This impedance was most likely chosen because most telephone systems use the 600-ohm balance line technique, and methods were already worked out when the broadcasting and recording industry got going. Also, many early and contemporary radio broadcast studios use the telephone system to transport the audio signal to the transmitting facilities. Control of audio level is all done at 600 ohms for either line amplifiers or attenuators. Such attenuators are either L, T, or π configurations to control the level and maintain the impedance match. A 600-ohm audio transportation scheme is shown in Fig. 3.1. The values for R_1 and R_2 versus loss, in decibels, for the constant-impedance attenuator pad in the figure are given in Table 3.1.

The standard signal reference level for a 600-ohm system is 1-mW dissipated by a 600-ohm load, which corresponds to 0 dBm. This is demonstrated in Fig. 3.2. The rms level across the 600-ohm load can be measured with a high-impedance true rms digital multimeter with

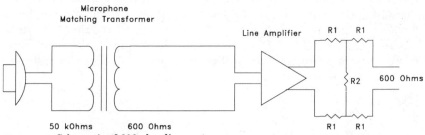

Figure 3.1 Schematic of 600-ohm line system.

TABLE 3-1   Common Values of Loss and Balance H Pad Resistances

dB	R_1 ohms	R_2 ohms
1	17.2	5200
3	51.3	1703
6	100	804
10	156	420
15	210	220
20	250	120

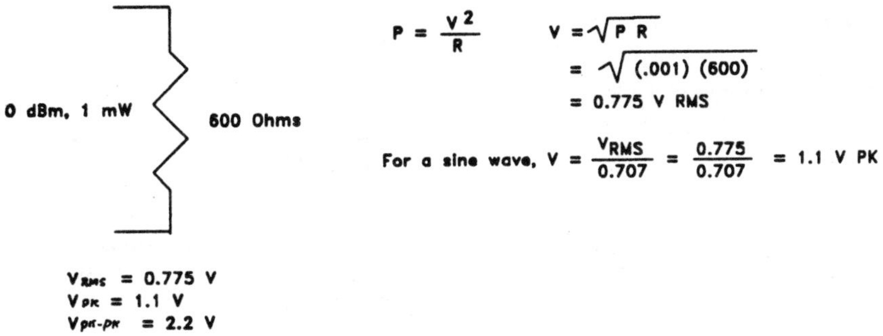

$$P = \frac{V^2}{R}$$

$$V = \sqrt{P R}$$
$$= \sqrt{(.001)(600)}$$
$$= 0.775 \text{ V RMS}$$

For a sine wave, $V = \dfrac{V_{RMS}}{0.707} = \dfrac{0.775}{0.707} = 1.1$ V PK

0 dBm, 1 mW   600 Ohms

$V_{RMS} = 0.775$ V
$V_{PK} = 1.1$ V
$V_{PK-PK} = 2.2$ V

Figure 3.2   0 dBm reference level.

the ability to measure accurately at the signal frequency. Also, a high-impedance oscilloscope can make either the peak-peak or peak measurement of voltage and then calculate the rms value. The oscilloscope method can also give an indication of any signal distortion.

Broadcast audio studios as well as most recording studios use what is known as a VU meter, which is essentially an audio signal volume indicator. Since the ear responds logarithmically to the acoustic power, the decibel is an important measurement of level. However, since the audio level in broadcast studios control the percentage of AM modulation and the frequency deviation ratio in the FM radio case a continuous reading of audio signal level is important. The scales on some VU meters also have units of modulation percentage on the instrument face. The VU meter is a specialized d'Arsonval meter movement with an accompanying copper oxide full-wave rectifier. The electromechanical construction changes the response characteristics for complicated waveshapes. Usually most multichannel audio mixing consoles contain one or more VU meters. A VU meter scale is shown in Fig. 3.3. VU meters are calibrated with 1 mW into a 600-ohm load, usually at 1000 Hz sine wave, which corresponds to 0 dBm and 0 VU. For a sine wave, one VU is nearly equivalent to 1 dB. The VU meter's dynamic characteristics attempt to duplicate the human ear's response to loudness. If a VU meter is placed across an impedance other

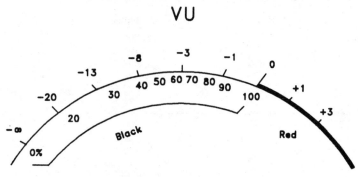

Figure 3.3   Example of a VU meter scale.

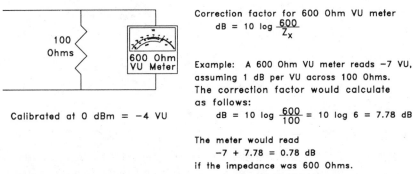

Figure 3.4   VU meter correction for impedances other than 600 ohms.

than 600 ohms, a correction factor has to be used. This procedure is shown in Fig. 3.4. Oftentimes a chart of correction factors versus impedance levels can be helpful if the meter is going to be used much at different impedances.

The electromechanical VU meter is still used today in many audio situations. However, the light-emitting diode (LED) bar graph display is becoming popular, mainly because it is easy to see and comparatively inexpensive. Its small size permits use of many VU/dB displays on audio consoles. This type of display used to monitor the level of several audio-frequency bands constitutes an audio spectrum analyzer. When level controls are included in the band of frequencies contained in such a spectrum analyzer, the audio band equalizer results. Even consumer "high-fidelity" audio systems contain either 7 or 9 band equalizers to tailor the sound to the listener's liking.

3.1.2   Audio power, gain measurements

Measurement of the power output of an audio power amplifier can be made with either an audio wattmeter or measuring the voltage and current through a resistive load often called a dummy load. The method is actually very similar, since many audio wattmeters have a self-contained accurate resistive load with a metering circuit built in. Essentially a signal source drives the amplifier to maximum power as measured on a wattmeter, and often an oscilloscope monitoring the load voltage can give an indication when distortion occurs. A simple power output test is shown in Fig. 3.5. Several manufacturers make audio power wattmeters with a usual accuracy of 2 percent. If no oscilloscope is available and since at the point distortion occurs the amplifier is at maximum output, a quick test would be to see if the amplifier can produce its rated output and to take a reading of its maximum output.

For line amplifiers, microphones, or phonograph preamplifiers, a measurement of undistorted gain is appropriate. Several audio test instruments used to measure signal distortion and noise also have a gain test function as well. A measurement of simple gain is to place a

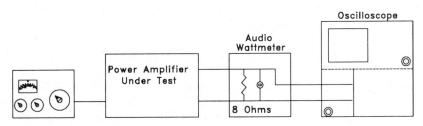

Variable Frequency
Signal Generator or
Function Generator
Sinewave Output

Adjust signal level of audio generator for the maximum allowable amplifier input. Increase the amplifier gain control until oscilloscope indicates distortion of the wave shape. Decrease gain for a distortionless signal. Read and record the wattmeter. Repeat for frequencies in the audio band.

Example: For a 25 watt amplfier, the oscilloscope indicated for a sinewave 40 V peak–peak (15.4 V RMS) just below the point of distortion. The wattmeter should read 29.6 W. This is confirmed by calculation:

40 V peak–to–peak = 20 V peak. (20) (0.707) = 15.4 V RMS

also, $P = \dfrac{V^2}{R} = \dfrac{(15.4)^2}{8} = 29.6$ W

Figure 3.5   Amplifier power output test.

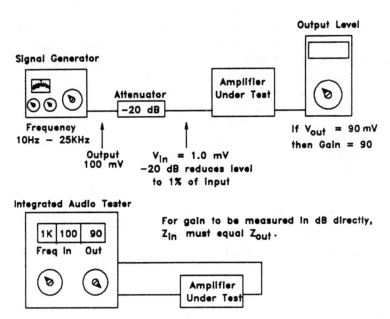

Figure 3.6 Gain-frequency amplifier test.

low-level signal on the input terminals and measure the input and output levels with an oscilloscope or an ac audio voltmeter. Such meters are commercially available with level ranges of about 100 μV to 100 V and a signal frequency range of 5 Hz to 1 MHz. Most audio voltmeters also have a dB scale as well, since many gain specifications are in decibels. A test setup for gain at various frequencies is given in Fig. 3.6. Essentially this method uses a function generator or audio signal generator as the input signal source and a wide-band-sensitive voltmeter across the output terminals. Some manufacturers place a generator, meter, and level attenuators in one package which is often referred to as an audio tester.

3.1.3 Amplifier distortion

Amplifier distortion is a major parameter that causes audio signal impairment. The two main types of distortion one reads in audio equipment specifications are total harmonic distortion and intermodulation. Both types of distortion result from poor linearity of the audio devices. The nonlinearity of the input/output characteristics, often referred to as the *transfer function,* causes the input waveshape to be altered at the output, hence causing distortion. Fourier analysis of the output-distorted waveshape will determine the changes in the frequency components between the input (pure) and output (distorted) signal.

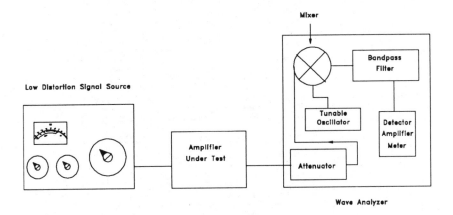

Example: if a 1000 Hz undistorted sine wave drives the amplifier under test and the wave analyzer measures a signal at 2000 Hz and 3000 Hz, the device has second and and third harmonic distortion.

Figure 3.7   Wave analyzer testing.

3.1.3.1 Total harmonic distortion.

A wave analyzer can be used to measure the amplitude of the frequency components contained in the output signal. This method of testing can give a measure of intermodulation and specific cases of harmonic distortion. An example of a simple harmonic distortion test using a manually tuned heterodyne wave analyzer is shown in Fig. 3.7. This instrument can be thought of as tuning a sharp filter through the audio spectrum and measuring and identifying the frequency components and distortion products. Also it may be considered as a sort of manually operated spectrum analyzer.

In some instances it may not be significant to identify the particular types of distortion but simply measure the total amount of harmonic distortion, known as the *total harmonic distortion* (THD). This specification is given as a percent of signal power, which increases with increase in power. For stereo hi-fi amplifiers the THD is generally given at full rated power output, that is, the worst case. An instrument used to measure THD is called a *distortion analyzer*. This instrument nulls or cancels out the fundamental test frequency and sums the remaining signal components that comprise the total distortion. Several manufacturers offer such instruments. Some are strictly manual and others are microprocessor-controlled with computer connections through IEEE 488 or RS 232 standard digital ports. Instruments with

Tek SG505 Oscillator   **Tek AA501A Distortion Analyzer**

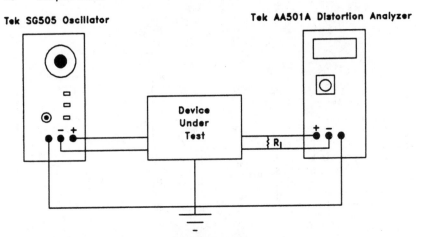

Figure 3.8   Distortion noise gain test configuration.

the ability to be set up, make measurements, and store data through computer control lend themselves particularly to production-line testing. The distortion analyzer test method requires a test signal generator to have an exceptionally pure, that is, low-distortion, sine wave signal. The instrument readout is usually in percent of signal level voltage, which of course is related to power. The Tektronix model AA501A Distortion Analyzer coupled with the companion SG505 Oscillator can automatically measure THD, THD + N (noise), and with option 01, intermodulation distortion. Measurements of gain and flatness can also be made. Many automatic features and a three-digit readout simplify the testing operation. A test setup is shown in Fig. 3.8. Measurements of THD can be made as low as 0.0025 percent, mainly due to the spectral purity of the SG505 oscillator (0.0003 percent THD typically). With option 02, THD measurements can be made in accordance with CCIR recommended standards (468-2) DIN (45405) as well as SMPTE.* If many repeated measurements are needed, such as continuous production-line testing, the Tektronix AA5001 programmable oscillator can automate the whole testing procedure through the IEEE standard 488 bus.† A personal computer with the same bus standards as the Tektronix PEP301 controller can program the instruments, process the data, and/or integrate other instruments in an overall testing program. When choosing equipment for a particular test, it must be remembered that the instruments have to be significantly better than the equipment being tested. When THD is at 3

---

*International Radio Consultative Committee (CCIR), International Deutsche Industrie Norm (DIN), European Society of Motion Picture & Television Engineers (SMPTE), USA.
†Institute of Electrical & Electronic Engineers.

percent, the signal level at that point is used in the signal-to-noise ratio. For instance, if the signal level is −5 dBm at 3 percent THD and the noise level of the device is −70 dBm, then the signal-to-noise ratio is 65 dB. Recall a ratio in logarithm form is numerator dB minus denominator dB.

3.1.3.2 Audio spectrum analysis and distortion testing. In more recent years the spectrum analyzer has become fully developed and hence more cost-effective, although even today still quite expensive compared to other types of instrumentation. This instrument presents a visual picture of the signal frequency level and the distortion component levels with a significantly large dynamic range. Essentially the spectrum analyzer acts very much like an automatic wave analyzer with automatic frequency tuning and sweep display. Digital circuitry similar to that used in digital oscilloscopes (DSO) is used in many state-of-the-art spectrum analyzers. The ability to store measurements and make comparisons with a screen (CRT) presentation can simplify the whole measurement procedure. The Tektronix 490 series can be programmed to transfer data through its digital bus to an external controller or personal computer. Most present-day spectrum analyzers have a large frequency range, large dynamic level ranges, a selection of bandwidth spread, and spectrum resolution. Some spectrum analyzers are special-purpose types that are able to examine a frequency range from very low frequencies, on the order of 10 Hz to 1 MHz. The Tektronix 497P covers a frequency range from 100 Hz to 7.16 Hz, which makes this instrument capable of handling a variety of measurements from audio to microwave. Before a decision is made to purchase expensive test equipment such as a spectrum analyzer, the first thing is to contact various manufacturers representatives, sales engineers, or application engineers and request some of the appropriate application bulletins. Also it may be advisable to rent such an instrument and try it out on a testing project. Many manufacturers will provide on-site demonstrations of the equipment as well as some training.

The three most important controls on any spectrum analyzer are the reference level, frequency, and the frequency span per screen division. The reference level sets the signal peak level to the top uppermost horizontal graticule line. The frequency sets the signal to be measured at the center of the screen or the left-hand-edge vertical graticule line. The span per division selects just how much of the signal width is to be displayed on a horizontal graticule division. Part of the frequency control is the resolution bandwidth (RBW) selection, which simply means the amount of bandwidth that sweeps through the signal. If the resolution RBW is too wide, then signals close to the

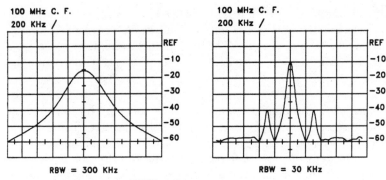

Figure 3.9   Differences in RBW settings.

center frequency will be obscured. If the RBW is made narrower, side band or close in signals will be resolved. Two screens of the same signal with just the RBW different are shown in Fig. 3.9. Also care must be exercised in selecting video or postdetection filters because they can cause amplitude errors as well. It goes without saying, to get optimum use and value from a sophisticated spectrum analyzer, the engineer should go through the appropriate learning experience. In other parts of this book use of the spectrum analyzer will be discussed. Harmonic distortion and THD can be measured with the spectrum analyzer as shown in Fig. 3.10. In this example the test signal is a 7-kHz sine wave with second and third harmonics at 14 kHz and 21 kHz, respectively. The second harmonic amplitude is 30 dB below the fundamental signal of 7 kHz and the third harmonic amplitude is down 35 dB. Since the spectrum analyzer gives a logarithmic display, the harmonics have to be reduced to voltage levels. To calculate the THD, one simply uses the square root of the sum of the squares technique, as shown in the figure. If many measurements are to be made, it might be advisable to make a graph of beat difference from the fundamental

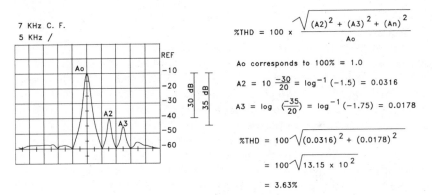

Figure 3.10   "THD" using the spectrum analyzer.

versus percent of harmonic distortion. Now the percent values merely have to be combined by the square root of sums of square method.

Intermodulation distortion is a type of distortion quite noticeable in audio work. A classical case of acoustical intermodulation distortion is in bells or chimes. The many harmonics generate beat notes that essentially blur the music. Measuring intermodulation distortion, commonly called intermod or simply IMD, uses what is known as a two-tone (frequency) test. The test signal generator provides test signals properly combined so as to not interact and cause distortion itself. By analyzing the input signal to the device under test, the spectrum analyzer will determine if the signals are distortion- (beat-) free. If, for example, the two tones (f_1) at 4 kHz and (f_2) at 5 kHz) are combined to drive an amplifier or radio modulator, then a spectrum analyzer can examine the output for the beats or intermodulation products caused by the signal distortion. Second-order intermodulation produces a beat at the sum-and-difference frequencies, namely 1 kHz and 9 kHz. Third-order IMD produces beats at $2f_1 + f_2$, $2f_1 - f_2$, $2f_2 + f_1$, $2f_2 - f_1$ (13 kHz, 3 kHz, 14 kHz, and 6 kHz, respectively). A spectrum analyzer screen for the IMD test is shown in Fig. 3.11.

Using the Tektronix AA501A and its companion signal generator SG505 with option 01 or 02 or the programmable model AA5001 and SG5010, the IMD testing can be made. These instruments have the capability of measuring IMD according to CCIR/DIN standards. The beat levels are read on the instrument's digital readout.

3.1.4 Audio noise causes and effects

All electronic devices generate noise. The noise level, measured as a voltage, depends on the bandwidth of the system. In the case of audio

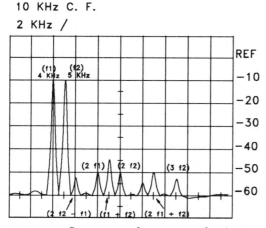

Figure 3.11 Spectrum analyzer screen showing IMD.

systems, the presence of noise is exhibited as hiss, which obscures the audio signal's high frequencies. Noise is generated usually at the input circuitry of an audio device such as an amplifier, and since audio amplifiers one way or another show up in many systems, noise gets generated and subsequently amplified. This buildup of noise can be very objectionable in audio program material. Noise generally builds up from the low frequency ranges to the high-frequency portion of the audio bandwidth. Low-frequency noise is usually power-line-related, that is, 60 Hz and/or 120 Hz, and is often referred to as *hum*. Also, this low-frequency hum can be very objectionable, consumes amplifier power, and obscures the bass notes of music. Faithful amplifiers and music-reproducing devices such as tape recorders, phonograph recordings, and compact disk players are the mainstay of the home music entertainment industry. Manufacturers and service repair people must have the capability of making noise and hum measurements in order to provide the high-quality performance their customers demand. Also, the sound systems in concert halls and recording studios usually have the capability to make noise and hum measurements.

3.1.4.1 Audio noise measurement.

The simplest method of measuring the noise and hum of an audio system requires a low-distortion, high-quality audio sine wave generator, an electronic audio voltmeter type (either analog or digital readout), some filters, and an oscilloscope. The equipment connection diagram is shown in Fig. 3.12. The lowpass filter passes the band of frequencies up to 150 Hz, which includes the power-line components of 60 Hz and 120 Hz that make up the hum products. The high-pass filter excludes the low frequencies, allowing the noise to pass through. Since white noise includes all frequencies,

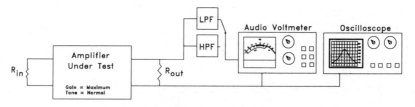

R_{in} matches the internal resistance of the amplifier.

R_{out} matches output resistance of amplifier.

HPF is a High-Pass Filter used in the Noise Test.

LPF is a Low-Pass Filter (150 Hz) used in the Hum Test.

The Oscilloscope reads noise level with HPF in circuit.

Set Horizontal Time Base and read hum level with LPF in circuit.

Figure 3.12   Noise and hum test.

the total noise measured will be slightly less. Recall that noise power depends on bandwidth and is often specified as decibels per hertz (dB/Hz).

Measurement of noise in audio systems can be easily made according to CCIR and DIN recommendations using the Tektronix AA501A/SG505 option 02 distortion analyzer setup. The noise weighting filters needed for CCIR/DIN recommendations are built-in and selectable.

The signal-to-noise ratio is a more often used and meaningful specification. When the difference between the signal level and noise level becomes close, the audio signal becomes objectionable. Since the ear's level response is logarithmic, the ratio of signal to noise, expressed in decibels, is often specified. It should be easily recalled that 60 dB represents a power ratio of a million to one. Therefore, a S/N ratio of 60 dB is quite good. A simple straightforward test for finding the signal-to-noise ratio is shown in Fig. 3.13. The narrow-band filter allows just the generated signal to pass through to the audio voltmeter, not signal plus noise. By removing the filter from the signal path and turning off the generator, the noise measured is just that generated by the device. By leaving the generator connected, the device has its input terminals' impedance matched. If it is suspected that the generator in its off condition is contributing some noise, a resistor (low-noise type) can be used to properly terminate the input of the device.

3.1.4.2 Noise, hum, and low frequency disturbances. Noise at audio frequencies is heard as hiss accompanying the music or speech. Since audio equipment covers the frequency range of 20 to 20,000 Hz, the noise

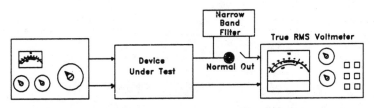

Low Residual Noise, Low Distortion
  Audio Signal Generator

Procedure: Set audio test signal at Narrow Band Filter's Center Frequency. Adjust device's output as measured on RMS Voltmeter through filter. Turn off Signal Generator and leave connected to maintain an input impedance match. Read noise voltage from device's normal output.

Example:
  Signal level V_s = 1 V
  Noise level V_n = 100 μV

$$\frac{S}{N} = \frac{1}{1 \times 10^{-4}}$$
$$\frac{S}{N} = 1 \times 10^4 = 40 \text{ dB}$$

Figure 3.13 Signal-to-noise radio test.

that falls in this range, usually above a few hundred hertz, impairs the sound quality. The frequency component below 150 Hz is characterized as hum and consists of power-line frequency components. Usually the power supply filter capacitors or other related power supply problems cause most of the hum. Also, if balanced audio (three-wire) input terminals are used, care in grounding the shields is important in controlling hum. If the hum is not generated in the power supply, then the most likely cause is direct hum pickup at the input terminals. Equipment using vacuum tubes can produce hum because the filament current is usually alternating current, and a gassey or shorted (cathode to filament) tube can cause hum. Some expensive older variety vacuum tube amplifiers had a dc power supply just for filament power to keep the hum level extremely low. Since the 60-Hz commercial power-line frequency falls in the audio range, it is the principal reason for hum generation. The hum specification is usually given as a percent of the output signal level. For high-quality audio amplifiers, the hum level should be less than 0.1 percent, which corresponds to 60 dB or greater.

Thermal noise is generated by resistors in the input stages of high-gain audio amplifiers. This is characterized by the well-known expression of the noise voltage, as described in Fig. 3.14, with a worked-out example of the input noise calculation. This input noise is actually the noise floor of the device, which means that it is as low as it's going to get. Usually the actual noise is larger, and the use of low-noise resistors and solid-state devices in the input stages can reduce the actual level closer to the noise floor. As was shown previously, the noise of a device is measured at the output terminals, but the cause is in the

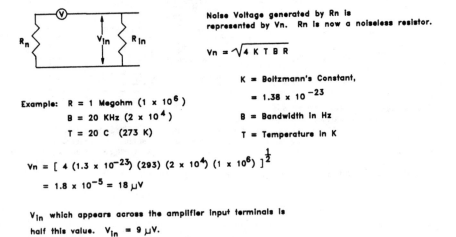

Noise Voltage generated by Rn is represented by Vn. Rn is now a noiseless resistor.

$$V_n = \sqrt{4 K T B R}$$

K = Boltzmann's Constant,

= 1.38×10^{-23}

B = Bandwidth in Hz

T = Temperature in K

Example: R = 1 Megohm (1×10^{6})

B = 20 KHz (2×10^{4})

T = 20 C (273 K)

$$V_n = [\, 4\, (1.3 \times 10^{-23})\, (293)\, (2 \times 10^{4})\, (1 \times 10^{6})\,]^{\frac{1}{2}}$$

$$= 1.8 \times 10^{-5} = 18\ \mu V$$

V_{in} which appears across the amplifier input terminals is half this value. $V_{in} = 9\ \mu V$.

Figure 3.14   Input noise voltage.

high-impedance input terminals. Microphones are usually low-impedance, and the cables are shielded for the long runs to the mixing amplifiers. However, at the amplifier the impedance is usually raised to a high value by matching transformers or a solid-state impedance transformation device. Examination of the equation for the noise voltage indicates that the noise power is proportional to the square of the noise voltage. When white noise of constant amplitude at all frequencies is considered, noise power increases with increasing the bandwidth. Therefore doubling the bandwidth doubles the noise power and increases the noise voltage by the square root of 2 (1.414); allowing the bandwidth to be larger than necessary allows more noise into the system. Using a filter that simulates the frequencies required for speech allows a band of noise known as *pink noise*. Audio engineers often use pink-noise-generating equipment to set up microphone systems in recording studios or concert halls. Noise generated in audio systems is a subject in itself and many books and technical papers have been written on this subject, some of which are listed in the bibliography at end of this book.

3.2 Crosstalk

Another interfering signal that causes problems in audio systems is crosstalk. This term came first from the telephone industry because background conversations were heard in the telephone receivers. Since telephone cables consist of many twisted pairs of copper wires, the capacitance existing between wires caused some intermingling of telephone conversations. Older type lead-covered cables with silk or fabric insulation on the copper conductors allowed varying degrees of crosstalk, mainly due to moisture entering the cable and thus decreasing the insulating qualities. This allowed a lower-resistance path coupling the conductors, together increasing crosstalk. Modern-day plastic-insulated copper wires have solved the telephone industry's crosstalk problems. However, the term is still used and it refers to any mixing of signals due to some means of cross-coupling. At audio frequencies, the coupling caused by capacitance between audio circuits is kept low by modern cables. At high gains and the concentrated wiring contained in audio switching panels and mixing boards, crosstalk can occur usually from a malfunctioning component.

3.2.1 Stereo separation

Some circuits exhibiting crosstalk are in stereo audio systems, where it is desirable to maintain separation between the left and right channels. Stereo circuits should have the ability to separate the left and right audio channels at least 30 dB. Many mixing audio panels in radio and television stations and recording studios have the ability to allow the left and right channels to be mixed or separated by the sound engineer's control. Unwanted pickup of signals from other channels is often caused by magnetic pickup. The twisting of the audio wire pairs causes some cancellation of the extraneous pickup. The outer metallic shield, either foil or wire braid or combination of both, protects the primary signal from unwanted signals. This shield should be well grounded at the amplifier input chassis.

The usual measurement parameter for crosstalk is the difference in signal level separation between the primary signal level and crosstalk signal level, measured in decibels. In the case of stereo, crosstalk is actually a measure of loss of stereo. The procedure for testing for crosstalk is to drive the right channel to full output level and measure the left channel output with the input properly terminated for any right channel ingress. This procedure is shown in Fig. 3.15.

3.2.1.1 Audio/stereo switching testing. Other systems susceptible to crosstalk and/or stereo separation are audio switching systems, which are used in a variety of situations. Some of these switching systems are used to select audio feed circuits from various studios to the main production area. Studio microphone or audio lines are often preamplified and the impedance lowered, in many instances, to 600 ohms balanced. Then the studio lines are switched by a bank or matrix to a production area consisting of the master mix board. When caution has been observed to shield these lines against noise, hum, and cross-coupling, the high concentration inside the switching system can defeat the whole process by allowing crosstalk, that is, cross-coupling of the audio signals. The switching system usually has active inputs with signals at the input terminals coupled to the contact

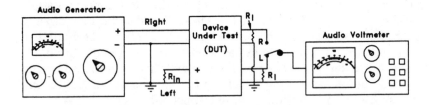

Procedure: Adjust generator for rated input level of DUT for the right channel. Measure the rated maximum output level as read on the meter for the right channel (switch at position R). Read right channel leakage on meter (switch at L). Repeat for several frequencies. Repeat for the left channel.

Figure 3.15   Stereo crosstalk test.

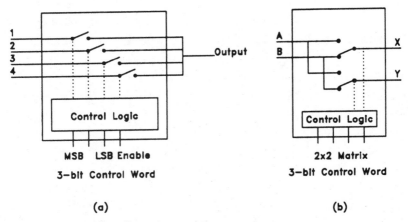

Figure 3.16   Switch configurations.

point. The switch selects various inputs to a common output, an example of which is shown in Fig. 3.16*a*. Matrix switches operate like a telephone crossbar switch, which has many inputs and outputs that can be switched either under a manual or an automatic control system such as a computer or microprocessor. A switch system of this type is shown in Fig. 3.16*b*. Testing a switch for crosstalk uses a technique similar to the one previously discussed for stereo separation, simply to drive an input and test the outputs. To properly test a multiple switch such as that illustrated in Fig. 3.16*a* and *b*, one input then two, three, and so forth is tested, while measuring output one, two, and three, and so forth. The procedure for this technique is shown in Fig. 3.17. All combinations of the inputs driven and the outputs tested are necessary to properly test the switch. For large-matrix switches, another computer-controlled switch may be used to select the tested switch combinations for measurement. The computer that programs the switch to be tested can extract data from the bus-controlled oscilloscope or audio voltmeter. The bus-controlled Tektronix AA5001/ SG5010 distortion analyzer test setup can be utilized with the PEP301 controller, or a PC can pretty much automate the process if many large-matrix switches are to be tested. Since many switch systems used today are solid-state and include control logic which requires some form of power, a dc offset voltage may appear at the output or input terminals. Manufacturers of such switches give a specification for any dc offset. If this offset, however small, could cause a problem with any equipment connected to the input and output terminals, the magnitude of the offset voltage should be measured. All terminals should be measured using a sensitive dc voltmeter (DVM) for any dc value while terminating various input and output terminals and turn-

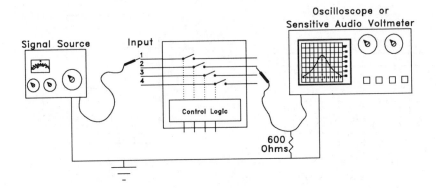

Procedure: Signal source (with proper impedance match) drives switch at maximum specified signal level. Cross—coupled signals are measured with input 1 active and then repeated for other inputs active alone then combined while testing the open (not connected) output terminals.

Example: The applied signal to Terminal 1 is 1 V. With all switches open, the signal output at A is 0.1 mV and 0.05 mV at the others. The isolation at A is 0.1×10^{-3}, or dB = 20 log (1×10^{-4}) = −80 dB.

The isolation of the other terminals is 5×10^{-5}, or 94 dB.

Figure 3.17 Switch crosstalk test.

ing the switch contacts on and off. It must be remembered that the switching system is essentially a device in the signal path and should be tested for gain (unity) or loss as well as signal distortion. If switching time is important, then it may be necessary to test for this parameter.

3.2.1.2 Shielding effectiveness of cable.

The shielding effectiveness of cabling in audio systems, particularly for long runs, can be a problem. Most manufacturers of shielded audio cables or coaxial audio cables give a shielding effectiveness specification, usually in percent. This simply means the signal voltage can be measured through the shield so many percent below the signal level carried by the cable. There is no simple method for measuring the shielding effectiveness in most industrial laboratories. Manufacturers usually have custom-built chambers where a sample of cable is inserted. This sample is properly terminated with a shielded termination and makes up the center conductor of a larger coaxial piece of cable that is usually made from a metallic tube. This larger coaxial cable segment is terminated at one end and the instrument measuring shielding effectiveness at the other end. When the center conductor of the sample is driven, the only signal measured by the detecting instrument is what gets through the

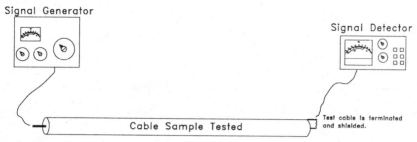

Figure 3.18 Shielding effectiveness test chamber.

shield of the sample. This method is illustrated in Fig. 3.18. Care should be taken to make sure terminations are tight and secure. This method is used to make comparison tests between cables, not so much for exact measurements.

If it is suspected that a cable is not shielded enough, it is not too difficult or expensive to replace it with a better shielded variety. Most manufacturers offer a choice of many different cable types. If the shield specification seems questionable, most manufacturers will supply more data on the particular tests used to obtain the shielding effectiveness value.

Proper grounding and bonding methods should be observed to keep common mode signals to a minimum. Most common mode signals are of the power-line type and hence are easily recognized. Most sensitive instruments use a two-wire connection with a separate ground input connection feeding a differential input amplifier. Instruments of this type cover such applications in the operating manual. If the shield of a connecting cable becomes part of a ground loop, then hum can get mixed with the signal. An example of a ground loop is shown in Fig. 3.19. The signal source could be a low-output magnetic phonograph cartridge, and the ground loop could consist of the metallic phonograph base plate and the amplifier chassis. The shielded wire connecting the cartridge terminals to the amplifier input terminals can carry the common mode voltage between the two chassis grounds. The cure is to provide a bonding wire between the two chassises to essentially short circuit the common mode voltage source.

3.2.1.3 Audio equipment limits and standards.

Audio equipment should perform well enough so that there are no discernible acoustical impairments. All audio program storage devices, such as phonograph records, tapes, and compact disks contribute noise, signal distortions, hum, and low-frequency disturbances to the audio electrical signal. These factors have to be measured to the degree that any value loss will not be heard. Some people are able to detect small variations in

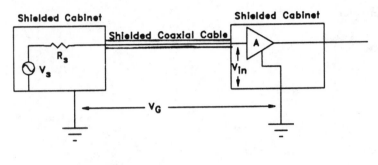

$$V_{In} = V_s + V_G$$

V_G is voltage between ground points.

Mixes with V_s and flows through shield of connected coax cable.

Figure 3.19   Ground loop problem.

noise, distortion, and frequency response, and hence have been said to have "golden ears." Most of us have to rely on making quantitive measurements to analyze the causes and cures of such audio signal impairments.

In general noise should be limited to at least 50 dB below the signal level and typically on the order of 60 dB. Total harmonic distortion should be less than 1 percent and for reasonably priced audio power amplifiers usually runs between ¼ and ½ percent, which amounts to 52 and 46 dB, respectively. This percentage is measured on a voltage scale with the device running at full rated output power. Hum and low-frequency distortions should be also 50 dB or better below signal level. Audio crosstalk should be kept to at least 50 dB or better, and most switching systems, even the solid-state variety, are able to typically provide better than 60 dB. Stereo separation is often found to be on the order of 25 to 30 dB. This is more than adequate, since room acoustics and placement of loudspeakers mix the left- and right-channel sound. The 25 to 30 dB separation gives a good stereo effect when listening with stereo earphones. Binaural sound requires more separation between left and right channels and is usually listened with stereo phones only.

3.3   Home Audio Equipment

Much audio programming involves the use of phonograph records; tape recorders, either cassette or reel-to-reel; and lately the optical compact disk. Today the popularity of the phonograph seems to be declining, although many people have a sizable investment in records. The tape recorder/player and/or the cassette tape recorder/player is the one method that allows the general public to record audio program material. The compact disc, like the phonograph, allows playback only at this time. It is expected that optical disk recording devices will become commercially available at reasonable prices in the foreseeable future. All of these program storage systems contribute some form of signal impairment characteristic to the particular method of recording and playback. At this time the compact disk player indeed does have the greatest signal dynamic range, excellent frequency response, and extremely low noise, which makes it the player of choice. All of these methods employ electric motors, belts, pulleys, and so on, to either spin the record or move magnetic tape across the heads. It is this electromechanical system that contributes mainly to any audio signal degradation.

3.3.1 Audio tape recording

Audio tape recorders are used by radio broadcast stations, television stations, concert halls, recording studios, and in the home and automobile as well. Tape recorders are either reel-to-reel, cartridge, or cassette type, and all have an electromechanical system to shuttle the tape across the record/play heads. The broadcast industry, recording studios, and concert halls still use reel-to-reel machines mainly because of their high-quality long-play characteristics and several recording tracks. Radio broadcast stations often use cartridge machines to insert commercials and/or station breaks. The cassette recorder has wide use in the home, automobile, portable recorders as well as dictating devices.

3.3.1.1 Tape wow and flutter testing. Tape recording equipment exhibits a unique form of audio signal impairment called *wow and flutter*. The term came from the effect of speed variations occuring in handling the magnetic tape. *Wow* refers to slow variations in pitch caused by either a defective motor, mechanical binding in the tape moving mechanism, or an out-of-round capstan or slipping pinch roller. *Flutter* refers to tape speed variations at a higher rate than wow. Wow refers to variations below 10 Hz, while flutter corresponds to variations above 10 Hz. Measurements are usually performed according to DIN (European) and or CCIR (International) recommended practices, using a wow or flutter meter and a precision test tape. Reasonably good quality tape machines mainly exhibit flutter, whereas phonograph equipment exhibits wow and flutter. The method of testing for tape recorder flutter is shown in Fig. 3.20. Another method is to use a test tape instead of recording and playing back the precision test tone signal. Essentially the precision signal is contained on the test tape and

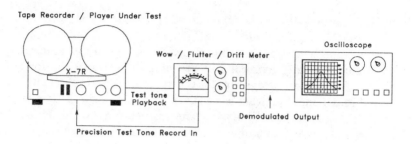

Procedure: The Test Tone (typically 3150 Hz) is recorded on tape and played back through the Wow/Flutter/Drift Meter. The Meter components modulate the Test Tone.

Normal record level, thus playback level is −10 VU.

Filter	Test	
		Analysis by the Meter is done by demodulating the reference signal as recorded by the machine. Comparison with the precision Test Tone using various filters to perform the analysis.
0.5 − 250 Hz	Wow, Flutter, Drift	
0.5 − 20 Hz	Wow, Flutter	
250 Hz − 5 KHz	Scrape/Flutter	

Figure 3.20 Wow, flutter, and drift test.

is played back on the recorder and analyzed by the wow and flutter meter. The oscilloscope connected to the demodulated output of the meter allows visual examination of the offending signals while diagnostic maintenance is being done on the equipment. The measurement of drift is shown on a separate meter in some commercially available wow-flutter meters. The drift parameter has a much slower variation in demodulated signal than wow and usually varies over several seconds. Most wow and flutter meters can measure in the range of .0025 to 10 percent (awful) at signal levels of as low as 0.5 mV. The value of wow or flutter can be in average, peak, or rms values to conform to DIN or CCIR standards. Most instruments of this type cost between $800 to $1500 and should not break most audio engineering budgets.

3.3.1.2 Audio tape recorder performance testing. Tape recorders should be able to record and reproduce the audio frequency band with sufficient accuracy to preserve musical pitch. Audio tape recorders incorporate a preemphasis system according to NAB recommended standards for tape equipment. The recording mode frequency response is tapered to accentuate the high-frequency portion of the audio frequency band. The inverse taper is incorporated in the playback mode to restore the overall record-playback band. A bias signal is also recorded on the magnetic tape as part of the recording process which is removed from the playback signal. To test the overall frequency re-

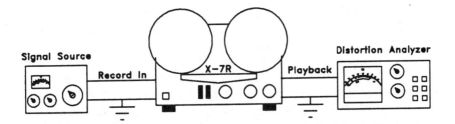

Procedure: Adjust Signal Source level and record gain to normal level on record VU meter (about −10 VU). Use Distortion Analyzer to measure playback level. A steps frequency is selected. Perform Harmonic Distortion tests with tape running and signal level reduced to 0.

Figure 3.21   Tape recorder frequency response, harmonic distortion, and noise test.

sponse of an audio tape recorder, the Tektronix SG505 low-distortion audio sine wave generator can be used as a signal source and the Tektronix AA501A distortion analyzer may be used in the configuration shown in Fig. 3.21. The setup can be used to measure the frequency response, harmonic distortion, and signal to noise using the features of the SG505 and AA501A. A low-distortion function generator with a sine wave function and an audio voltmeter can be used for the frequency response test. If a separate noise-weighting filter is available, this could be inserted in series with the voltmeter to make noise measurements provided the voltmeter has sufficient sensitivity. Some audio engineers prefer using a square wave signal as a signal source and an oscilloscope as an indicator. The type of distortion of the square wave at various audio frequencies is a measure of the overall response. The square wave is indeed a unique signal where the average, peak, and rms values are the same.

3.3.1.3 Tape recorder stereo separation.

Stereo separation is a feature that often needs testing since the left and right tracks are recorded together. Therefore two things can happen, there can be cross-track mixing and what is known as print-through can take place. If the recording level is set too high, distortion can result and the higher magnetic density can affect the tracks of the left and right channels which tend to magnetize through the tape as it is reeled on top of each succeeding layer of tape. This phenomenon can be realized as a low background noise that is heard during silent periods of program material. Print-through is sound heard ahead on the same track, since one track is essentially rolled on top of the other. To test for stereo separation, a simple test is to record a signal source on one track at full recording level while measuring the output on the other track with the same equipment used in the frequency response, distortion, and noise test. Since print-through is an inherent problem with the tape recording method, equipment performance is not really the culprit. If the recording equipment is not operated properly, print-through can be made worse. To actually measure print-through, short intervals (several seconds) of normal full-level program material on each track is recorded followed by several seconds of silence. During the silent period on playback, an oscilloscope or sensitive audio voltmeter can measure the amount of signal transferred from the recorded areas to the unrecorded areas of the tape. Again the same equipment configuration in the previous figure can be used. If it is desired to see if the print-through effect is objectionable, a listening test with stereo phones can be made by listening for sounds during the quiet periods on a recording.

3.3.2 Phonograph records

Phonograph records have been around for a long time. The first phonographs were purely acoustical. Thomas A. Edison developed the early method of recording and reproducing sound on plastic cylinders and later disks. The electromechanical phonograph cartridge was developed in later years and the disks were turned with electric motors instead of spring-energized mechanical motors. Early electric record players operated at 78 revolutions per minute and the stylus riding in the grove was metallic. The resulting sound was marginal and usually exhibited the sound of stylus or needle scratch so many of us remember with nostalgia. Later years brought the long-playing (LP) record, greatly improved phonograph cartridges, and styli. The slower record speed improved the recorded sound with reduced stylus noise. The phonograph or record player became a major source of home music entertainment even with all its inherent problems.

3.3.2.1 Phonograph record playing parameters. The successful playing of a phonograph record depends on a multitude of critical parameters. The played record should be able to reproduce the sound that was recorded on the master record. This means the recording speed should exactly match the playback speed so as to maintain proper music pitch. Low-mass, lightweight cartridges with diamond styli reduce playback scratch noise and improve the frequency response. By measuring the playback signal from a test recording, the slow speed changes that cause pitch variations can be measured by a wow and flutter meter just as in a tape recorder. Some commercial wow and flutter meter manufacturers have such test recordings available. The test setup configuration is essentially the same as for the tape re-

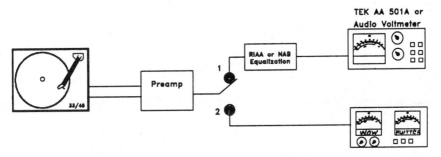

Set switch to 1 for Frequency Response (RIAA or NAB)
recording response curve.

Set switch to 2 for turntable Wow and Flutter.

Figure 3.22 Phonograph frequency response and wow and flutter test.

corder wow, flutter, and drift test except that a test recording is used
as the precision signal source as shown in Fig. 3.22. Since records are
made using either the NAB (National Association of Broadcasters),
RIAA (Record Industry Association of America) or the European DIN
equalization standards, the type of test record has to be taken into account for any frequency response test. Test records usually have
equalization information printed on the label near the spindle hole,
and if an equalization circuit is not available, the wow and flutter test
can be made, but an accurate frequency response test will be approx-
imated without the appropriate equalizer. The low-frequency wow and
drift test is often referred to as turntable rumble.

3.3.2.2 Turntable speed and speed testing. Since the turntable speed
sets the reproduced music pitch, long-term proper speed must be
maintained. Most high-quality turntables employ a precision motor
that is locked to the commercial power-line frequency of 60 Hz. Such
motors are called *synchronous motors* since their rotation speed is synchronized to the power-line frequency. To test for speed, a flash tube of
either zenon or neon energized by commercial power will flash at 60
times per second. Hence any mark placed on the turntable will be stationary visually if the turntable motor is indeed synchronized. A small
disk with various markings placed in bands and played on the rotating phonograph turntable illuminated by a neon lamp is commercially
available to test for proper speed. If the correct band appears stationary, the speed is correct. If other bands appear stationary, the speed
may be over or under the correct speed. A simple neon lamp circuit
driven by commercial power and the turntable speed test are shown in
Fig. 3.23a and b. Commercially available phototachometers can also
be used to measure turntable speed.

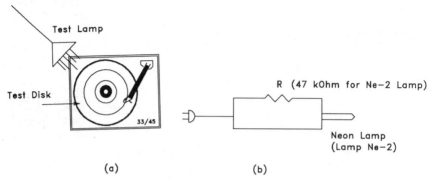

(a)   (b)

Figure 3.23   Strobe method of testing turntable speed.

3.3.3 The compact disk testing method

Compact disks (CDs), known also as compact audio disks, provide audio performance of significant proportions over records or tape. At the
present time tape is still the only audio medium that enables one to
record and playback in a single affordable device. The compact disk
essentially is a high-quality playback method with audio program material recorded on a special plastic disk. An audio signal is digitized
and the digital numbers corresponding to the audio signal are recorded by a laser on the disk as pits. A laser beam and optical system
read back the program material from the disk without touching the
disk. Thus the disk wear is extremely minimal, resulting in a low-
noise, high-quality playback system that outlasts either records or
tape. The dynamic range is about 90 dB and the signal-to-noise ratio
is nearly 90 dB. Therefore noise and response testing requires test
equipment and methods significantly better than the compact disk
players.

Since the compact disk has to be rotated, the speed is critical even
though the record/playback method is digital. Speed changes cause errors in reading the digital data, which result in a problem called *jitter*.
Commercial instruments are able to test for jitter by testing the digital clock pulses that are part of the recording process. Errors in the
shifting or changes in the pulses are given as shifting of these pulses.
Since these clock pulses fall in the range of 4.3 MHz, proper cable and
connectors should be used.

If test disks exhibit single-frequency-tone sections throughout the audio spectrum and quiet (blank) sections, the playback frequency response
and signal-to-noise ratio tests may be made using a wide-band sensitive audio voltmeter or the Tektronix AA501A distortion analyzer.

3.4 Summary

Audio engineers are mainly interested in three basic parameters: frequency response, noise, and distortion. The frequency response test can be as simple as driving the device under test with steps of single frequencies at constant amplitude throughout the audio range and measuring the output level, or by automatically sweeping the input signal source through the audio frequency range and measuring the output on either a synchronized oscilloscope or chart recorder connected to the audio voltmeter. Markers on the chart recorder signify the frequency steps. Digital storage oscilloscopes with the GPIB output bus can drive a digital printer with the frequency response trace.

Background noise can be measured by analyzing the signal content during silent periods of the recording. The phonograph record's principal noise source is stylus or needle scratch while magnetic tape noise is called tape hiss, which generally is less than record scratch. CDs exhibit the best noise (lowest) performance and since the signal-level dynamic range is so large, no equalization or pre- or deemphasis is needed.

Signal distortion is also an important audio specification. The most common specification is total harmonic distortion, which has been discussed as well as intermodulation distortion.

In the final analysis, the question remains is what sounds best? This fact gives rise to subjective testing or listening tests when a piece of equipment is actually used for both speech and music to a group of people. When carefully chosen music is played and people are asked questions on a printed sheet or press a group of buttons used to register an answer, an evaluation of audio equipment can be made. Surprisingly, quite slight variations in playback parameters show up in subjective listening tests.

Measurements at RF Frequencies

4.1 RF Communications and Equipment

Many modern communications systems have electrical power sources at frequencies ranging from the hundreds to thousands of megahertz. The reason is simply spectrum space for communications, radar, and navigational systems. The higher the frequency of electrical power that can be generated and used for such purposes, the more expanded is the spectrum space available for various uses. Most of us in the electrical and electronics industry know the effects and characteristics of high-frequency power and its ability to propagate its electric and magnetic fields through space. Measurements at the high and superhigh frequencies can present a problem. However, many instrument manufacturers have constantly improved and upgraded their equipment to make accurate measurements at the high frequencies. Many voltmeters, ammeters, and wattmeters have generally improved to the point that common measurements can be made at much higher frequencies than before.

4.1.1 RF electrical measurements

Several present-day digital multimeters can directly measure voltages to 1 MHz at true rms and sine waves to 10 MHz. Some manufacturers using special electrometer vacuum tubes and circuitry can directly measure true rms alternating voltages to better then 1 GHz provided the voltage magnitude is in the range of 25 mV. If higher voltages in this frequency range are desired, a resistive and properly shielded voltage divider can be utilized.

Measurement at radio frequencies of current may be taken both directly and indirectly. If the current is passed through a properly shielded shunting resistance, the voltage may be measured directly and the resulting current calculated. This method may be referred to as indirect. Current measured this way is frequency-limited by the voltmeter. High-frequency ammeters, often called RF ammeters, are of the thermocouple type and are usable up to about 50 MHz. Often these meters are used in antenna circuits to monitor the current drawn by the antenna. Antenna system deterioration or changes in tuning can be spotted by monitoring the transmission-line current feeding the antenna. Also using an RF power meter and knowing the circuit impedance both real and imaginary parts, the voltage and current can be calculated as well as the phase angle between voltage and current.

The measurement of power at high frequency is often more meaningful than simply voltage or current. Present-day thermistor power meters usually operate at 100 to 200 ohms impedance and are able to measure power at frequencies between 100 kHz to about 25 GHz at accuracies of 1 percent with power ranges between 0.1 mW to 25 W (– 70 to +44 dBm). Frequencies in these ranges are often referred to as microwave frequencies, since the wavelength is short and in the range of centimeters. As the frequencies increase, so do the measurement problems and of course the accompanying increase in instrument costs.

4.1.1.1 RF voltage. The choice of RF voltage instruments depends on the magnitude and frequency of the voltage to be measured. For instance, at 100 MHz, RF voltmeters are available that will make the measurement directly for voltages from 10 mV to 1000 V with an accuracy better than 0.25 percent. Also, many present-day oscilloscopes can make measurements of voltage up to 500 MHz, such as the Tektronix 7904A or the new 11401. To extend the frequency range to microwave applications, sampling-head plug-in modules are available for several oscilloscopes. The sampling head-method depends on the fact that the frequency is continuous and the waveshape is sampled at various points on the incoming wave. Accuracies of voltage levels on the order of 0.25 percent at 1 GHz are possible. Such oscilloscopes naturally are extremely expensive but give information on waveshape (distortion) frequency and amplitude. Most voltmeters and oscilloscopes are usually fitted with appropriate high-frequency probes and cables with proper insulation and shielding. Since instrument cable lengths are usually short, from 1 to 3 m or 3 to 10 ft, care must be taken that impedances are properly matched to avoid standing waves and the accompanying error. Fifty ohms impedance is pretty much standard for high-frequency instruments and test cables often are made with RG58 50-ohm cable fitted with standard 50-ohm BNC type connectors. Instruments using such test cables and connectors are operated at or below 1 GHz. Higher-frequency equipment uses larger-type 50-ohm cable with type N connectors and operates to several gigahertz. Higher into the microwave-frequency spectrum instrumentation uses either rigid coaxial cables or wave guide.

4.1.1.2 RF power measurements.

As stated before, at microwave frequencies the measurement of power is indeed more meaningful because unlike voltage and current, the level of power in a lossless transmission line is nearly constant. Measurement of power at the terminal devices allows measurements of gain and loss between source and load. Power meters are available with either analog or digital readouts and many may also be programmed by a digital computer or microcontroller. Such an arrangement also has the benefit of the ability to extract digital data from the instruments and print either lists or plot curves of the power measured. The heart of most power meters is the sensing element, which essentially responds to the microwave-frequency power that is impressed on it and generates a dc analog signal proportional to the power. The sensing elements may be either special diodes, thermistors, or thermocouples. Sensors manufactured by Hewlett Packard are usually paired with several of the company's meters with a choice of various connecting cables. A calibration record is included with each sending element traceable to the National Bureau of Standards Laboratory. Sensing elements are available in specific frequency ranges and power level ranges at the usual 50- or 75-ohm impedances.

The circuit configuration of making a measurement of microwave power from a source is shown in Fig. 4.1. Caution should be taken to make tight connections so as to avoid standing waves which can affect the measurement. All of the devices should be properly impedance-matched for the same reason.

Another method of making power measurement is by using a waveguide mixer such as the Tektronix series WM782. This device is similar to the sensor. However, the microwave energy is down-

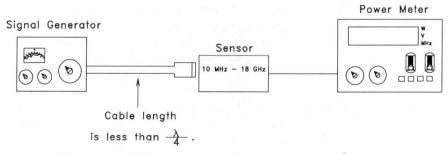

Figure 4.1 Measurement of microwave power using a power meter and sensor.

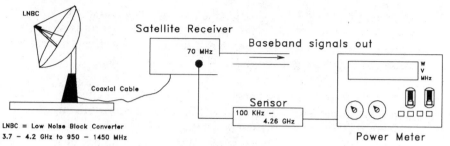

Figure 4.2 Measurement of C/N at 70 MHz IF for a satellite receiver.

converted to a lower frequency with an associated calibrated power loss, with the loss versus frequency characteristic printed directly on the device. Essentially this device is a companion piece to the Tektronix High-Performance Microwave Spectrum Analyzer, model 2782. The mixer's local oscillator is derived from the 2782. Thus only one cable connection between the instrument and the mixer is needed. The WM782 extends the frequency range of the 2782 to 220 to 325 GHz at a sensitivity of −75 dBm. By analyzing the spectrum, the distribution of power versus frequency can easily be made. This tells a lot more about the signal than just a quantitive power measurement.

However, such a measurement of power with a power meter still provides power gain and loss information. A method of measuring relative carrier power-to-noise for a satellite receiving antenna is shown in Fig. 4.2. The measurement is performed at the receiver IF frequency of the usual 70 MHz. The receiving antenna is peaked on the satellite transponder for both azimuth and elevation as well as any polarization, and the signal level is recorded on the power meter. Then the antenna is moved off the satellite a few degrees and the noise measured in dBm. The difference between the signal power in dBm and the noise power in dBm is the carrier- (IF) to-noise ratio in decibels. This method is often used to aid in the pointing and polarization adjustments by maximizing the carrier-to-noise ratio. The manufacturer's catalog information contains the specifications of the various sensors and companion meters to perform a variety of power measurements. However, most manufacturers of instruments have application engineers available to aid prospective users of the various testing and equipment alternatives with the respective costs.

4.1.1.3 RF high-power testing and precautions. High power levels at radio frequencies from a few hundred kilohertz on up to microwave frequencies can indeed be injurious. The original diathermy machines used by many physicians operated in the region of several megahertz

TABLE 4.1   ANSI Limits of Safe Electromagnetic Radiation

Frequency	Limit, mW/cm^2
300 kHz–3 MHz	100
25 MHz–300 MHz	1
2 GHz–100 GHz	5
5 MHz	50
10 MHz	10
600 MHz	2

and heated from the inside to outside. This is essentially how microwave ovens operate. Many people working around radio transmitters at powers greater than 500 W have received some RF burns on their hands or fingers. Healing of such burns usually takes much longer than ordinary cuts. The safe level for electromagnetic (radio) radiation varies with frequency and the limits (in microwatt per square centimeter) not to be exceeded is given in Table 4.1. This is the specification set by the American National Standards Institute (ANSI).

4.1.2   Special Purpose RF Testing Equipment

High-frequency measurements, particularly in the microwave region, often require special-purpose instruments. The measurement of high power levels at microwave frequencies utilizes what is known as a calorimeter or in common terms, a water load. This method is essentially a laboratory method and often is constructed and designed for a specific piece of equipment or use. Such equipment is not commercially available from instrument manufacturers' standard product lines.

The use of commercially manufactured RF bridges, return loss bridges, precision directional couplers, and slotted transmission lines make possible a variety of RF measurements.

4.1.2.1   The directional coupler test port. The directional coupler is a particularly interesting and simple device that is often used to measure and monitor power levels in a transmission line circuit. Directional couplers are made in a variety of impedances, the most common at 50 or 75 ohms. Power-handling capacities and degrees of directivity are other important specifications. Essentially a direction coupler is a device which when placed in a transmission line, responds to a wave traveling in a specific direction, usually the forward wave. Also, directional couplers may respond to the reflected wave as well, and when used in combination, forward and reflected power can be measured. The directional coupler element is often integrated into a specific instrument. The instrument has a meter circuit that measures a voltage

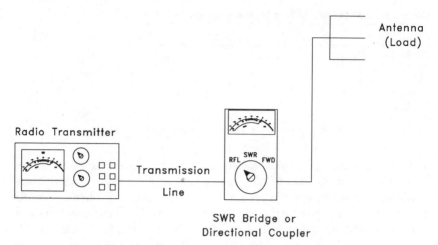

Figure 4.3   Measurement of SWR.

which is proportional to the current flowing in the transmission line through the directional coupler. The bridge type of directional coupler is commonly referred to as an SWR (standing wave ratio) bridge. The SWR bridge readout is a metering circuit that is calibrated in SWR. The circuit connections for both types of instruments are the same and are shown in Fig. 4.3. Essentially the instrument is placed in the line (in series) with the transmitter and the load (antenna). As a refresher, the relationship between SWR, impedance, and return loss is given in Fig. 4.4. It should be remembered that a standing wave caused by a

Voltage Standing Wave Ratio is defined as

$$VSWR = \frac{Vl - Vr}{Vl + Vr}$$

$$= \frac{1 + K}{1 - K}$$

Vl = Incident Voltage

Vr = Reflected Voltage

K = Reflection Coefficient

$$\text{Therefore } K = \frac{VSWR - 1}{VSWR + 1}$$

$$K = \frac{Z_1 - Zo}{Z_1 + Zo}$$

Zo = Characteristic Impedance

Z_1 = Load Impedance

$$\text{Return Loss} = 20 \log \frac{1}{K}$$

Figure 4.4   Relationship of VSWR, reflection coefficient and return loss.

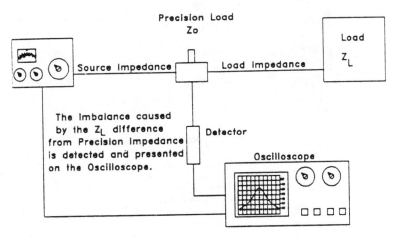

Figure 4.5 Swept return loss test.

reflected portion of the transmitted power is due to an impedance mis-match, usually at the load. This impedance, if accurately matched from source to load, will give a reflection coefficient of zero, a standing wave ratio of 1:1, and a return loss approaching infinity. A return loss bridge can measure the return loss in decibels and the reflection coefficient can be calculated. This device employs a signal source and a precision-terminating impedance. If the signal source is a sweep generator and the receiver a detector connected to an oscilloscope, the return loss over a band of frequencies can be determined. In some instrument setups, the signal source can be an RF noise generator and the detector an RF signal-level meter. The signal-level meter is in essence a frequency-selective voltmeter with variable gain controlled by an attenuator. Therefore, return loss can be measured by the meter at selected frequencies as tuned by the tuning control. The test for return loss is shown in Figs. 4.5 and 4.6. An instrument is available that contains the noise source and the bridge, so all that is needed is a signal-level meter to give the return loss value directly.

4.1.2.2 Slotted transmission line testing of Voltage Standing Wave Ratio (VSWR). The slotted transmission line can provide an accurate test for terminating impedance by measuring the precise VSWR. Slotted transmission lines are available for high-frequency coaxial configurations as well as microwave waveguides. A slot is longitudinally cut along the wall of a transmission line and a detector probe is mounted on a carriage and inserted in the slot. As the detector probe is moved along the slot, the amplitude of the RF level is measured. Consecutive peaks (maximums) and the valleys (minimums) give the VSWR. A

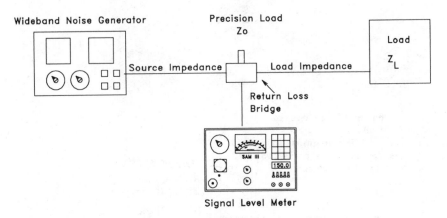

Figure 4.6 Return loss using a wideband noise generator.

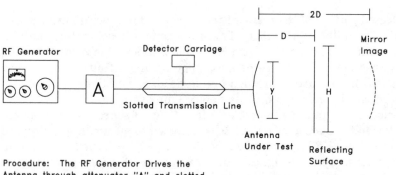

Procedure: The RF Generator Drives the Antenna through attenuator "A" and slotted line. The Antenna is impedance matched to the system. The reflecting surface is placed in front of Antenna causing VSWR "S" to change as measured by slotted line. Attenuator "A" isolates the RF Generator from reflections. The Antenna Gain is given by

$$\text{Gain} = G = \frac{4\pi \, 2D}{\lambda} \, \frac{(s-1)}{(s+1)}$$

y = 5 m (3 GHz)

3 GHz has a wavelength of 0.1 m

$$2D = \frac{2\,(5)^2}{0.1} = 500 \text{ m}, \quad D = 250 \text{ m}$$

$$H = \frac{2\,D\,\lambda}{y} = \frac{(500)\,(0.1)}{5} = 10 \text{ m (should be a 10 m square surface)}$$

For S = 2:1, $G = \dfrac{4\,(3.14)\,(500)}{0.1} \, \dfrac{2-1}{2+1} = 20933.3$

y = diameter of Antenna
H = height of reflective surface
D = distance between Antenna and reflector
A = Attenuator
S = VSWR
λ = wavelength
Note: 2D > 2y

$$G_{dB} = 10 \log (20933.3) = 43 \text{ dB}$$

Figure 4.7 Measurement of antenna gain.

slotted line is considered a laboratory piece of equipment and is used to measure antenna impedance as well as RF component impedance to a high degree of accuracy. Use of a slotted line in a test for antenna gain where the slotted line measures the VSWR is shown in Fig. 4.7. The reflecting surface H has to be for a square surface large enough to reflect all of the energy transmitted by the antenna with an aperture of y. The antenna with the best gain will transmit most of the power to the reflector, which in turn will reflect most of the power back to the antenna. Thus the worse the VSWR, the higher the antenna gain. For a VSWR of 5:1 then the antenna gain becomes 46 dB.

4.1.2.3 Frequency measurements.

Measurements of frequency to a high degree of accuracy are extremely important. High-frequency signals radiate into space, causing interference with duly-licensed (FCC) and authorized users. The users of radiated energy are the network broadcasters, other members of the telecommunications industry, plus the military and government agencies. The users of such spectrum space must maintain frequency accuracy as specified and controlled by the Federal Communication Commission (FCC) in the United States. Such limits are published by the FCC as well as the National Association of Broadcasters (NAB), the Institute of Electrical Electronic Engineers (IEEE), and many other professional societies. In recent years the measurement of frequency to a high degree of accuracy has been made possible by the ability to measure time by extremely accurate means. The atomic clock is controlled by a cesium atomic beam which controls an oscillator to a high degree of accuracy ($\pm 5 \times 10^{-12}$ parts or seconds at 20°C) and a long-term stability of about $\pm 3 \times 10^{-12}$ parts or seconds. A rubidium-type atomic frequency standard has a typical accuracy of about $\pm 2 \times 10^{-12}$ parts, but the long-term drift rate is 1×10^{-11} parts or seconds on a monthly basis.

The precision quartz oscillator does not have such accuracy and long-term stability, but it still offers a frequency standard of sufficient accuracy for many applications. The output frequency can be adjusted to within $\pm 5 \times 10^{-8}$ parts. After a suitable warmup, most quartz frequency standards are at 5 parts in 10^{12}.

These frequency standards usually offer test frequencies at 100 kHz, 1 MHz, 5 MHz, and 10 MHz, which are used to check frequency counter operation. To calibrate most frequency counters, the frequency standard feeds a precise signal to a frequency counter. The time base of the counter is adjusted to read the precise frequency. This test should only be performed after both instruments have been powered on for 30 minutes.

The National Bureau of Standards transmits radio signals at the precise carrier frequencies of 5, 10, and 15 MHz. A cesium beam fre-

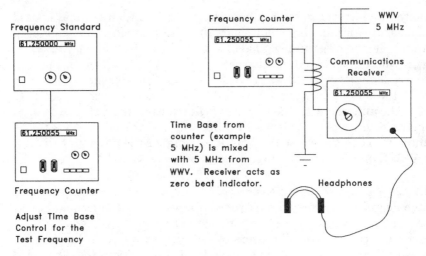

Figure 4.8 Frequency counter calibration methods.

quency standard controls the transmitted carrier frequency. Thus this received signal can also act as a frequency standard to calibrate a frequency counter. The two methods of calibrating a frequency counter are shown in Fig. 4.8. Since most industrial laboratories do not have such frequency standards, many independent testing laboratories are available to perform such calibration services. One manufacturer of a frequency counter has a removable time base module that can be exchanged with a factory-calibrated one. This can be done on a rotational basis every year or so and the calibration is traceable to NBS standards.

4.2   RF Broadcast Equipment

High-frequency equipment, even today, is mostly used by the broadcast industry, which consists of AM and FM radio services and of course television. Other uses are industrial induction heating for sealing and curing plastics and metals, and in medicine for heating various parts of the body for treatment or diagnostic purposes. The final RF output is usually a power-amplifying device, and although some are solid-state, most use vacuum tubes. RF oscillators are used to generate an RF signal followed by either amplifiers or frequency-multiplying amplifiers drive the output power amplifier. In the case of radio and TV broadcasting transmitters, modulating equipment is used to place the intelligent signal on the RF carrier.

4.2.1   FCC rules for broadcast

Since the broadcasting industry is under strict rules and regulations administered by the FCC, broadcasting transmission equipment has to be constantly measured and monitored with the results recorded in a logbook with specific requirements. Frequency accuracy and stability, power input to the final amplifier stage, radiated power output, and percentage or index of modulation are some of the important parameters to be monitored and logged. Each broadcast station has to be duly licensed by the FCC and proof of personnel technical competence has to be proved. Violation of rules and regulations can cause either fines or forfeiture of license or both. Hence most broadcast transmitters have built-in metering and monitoring circuits to provide the required information.

4.2.1.1 Measurement of transmitter power. The power output is often monitored by measuring the final amplifier current and voltage as well as the RF antenna current. The meter is often of the thermocouple type, and by measuring the true rms current value and since the antenna is tuned properly for a resistive load, the power can be calculated. Also, many antenna structures contain small-loop antennas coupled to a diode voltmeter circuit which supplies a dc output voltage proportioned to the induced RF current in the loop.

Transmitter proof of performance, often with a partial proof once a year, consists of measuring the radiated field-strength around the antenna site. Constant field-strength contours are plotted in microvolts per meter to prove that the transmitter is operating within its licensed geographical area. This instrument is actually a superheterodyne tuned receiver with a calibrated-loop antenna and metering circuit calibrated in microvolts per meter. Such instruments are commercially available and manufacturers usually have calibration services available. A block diagram of a fairly typical AM radio transmitter power-monitoring system is shown in Fig. 4.9.

4.2.1.2 Modulation testing. The modulation parameter is of extreme importance because overmodulation of either the FM or AM band can cause out-of-frequency operation, which is a violation of FCC rules. Many standard broadcast AM transmitters have a percentage of modulation monitor at the transmitter site. The broadcast engineer in charge should enter the measured modulation percentage into the log along with the power measurements. A simple test of AM modulation using a low-distortion audio signal generator TEKSG-505 and an oscilloscope with X and Y axis available such as TEK2205 will give the traditional trapezoidal pattern illustrated in Fig. 4.10.

Also, AM modulation can be estimated using the same equipment, but only examining the modulated RF output sample. This method is shown in Fig. 4.11.

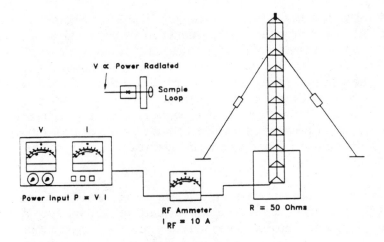

$$P_{out} = I^2 R = (10)^2 (50) = 5000 \text{ Watts}$$

If the antenna has a gain of 3 dB then
the effective radiated power is
10,000 Watts (5000 x 2)

Figure 4.9   Radio transmitter power test.

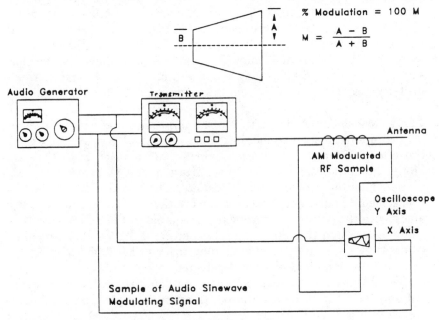

Figure 4.10   Amplitude modulation measurement.

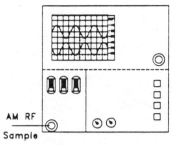

% Modulation = 100 M

$$M = \frac{A - B}{A + B}$$

Sweep Rate = $\frac{1}{2}$ Audio Rate

Figure 4.11   Oscilloscope method of AM Test.

4.2.1.3 Carrier frequency tests. The measurement of carrier frequency is always important. The most accurate means of measuring frequency by today's standards is a high-quality properly calibrated frequency counter. Most modern transmitters used in the broadcast industry also contain frequency-monitoring equipment. Since frequency counters actually measure the number of cycles in a precise time frame of one second, the presence of modulation can fool the instrument. Therefore the frequency should be monitored at a point before modulation takes place so as to avoid the problem.

4.2.2 The television transmitter

One of the more complicated and involved types of transmission equipment is the television broadcast transmitter. Since the television signal contains both a video and audio component, the transmission equipment is often separated until the final output stage. For the standard video NTSC signal, AM-type modulation is used with the lower sideband removed by a vestigial sideband filter. The audio signal modulates a 4.5 MHz subcarrier in the FM mode to provide high-quality audio television reception. The video signal has to provide both the vertical and horizontal picture-synchronizing signal as well as the brightness and color information. Parameters that require monitoring and logging are power output, both video and audio modulation levels, visual/white visual sync to picture, audio modulation level, and frequency. Careful monitoring of the television transmitter's operation is required by the FCC, and of course is necessary to provide high-quality television service.

4.2.2.1 NTSC video modulation. The signal waveform for the NTSC (National Television Study Committee) video modulating system is shown in Fig. 4.12. The television receiver recovers (demodulates) this

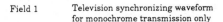

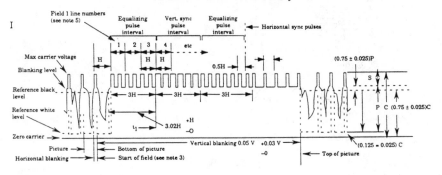

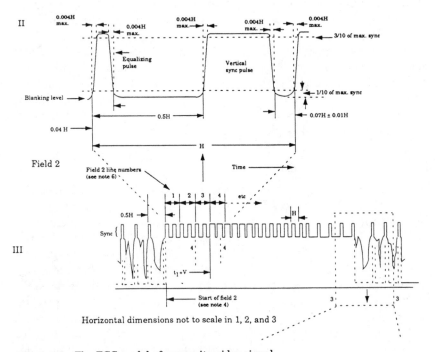

Horizontal dimensions not to scale in 1, 2, and 3

Figure 4.12 The FCC model of composite video signal.

signal from the transmitted RF carrier and produces a color image on the receiver screen. Since there are many books written on color television, only the methods of testing the purity of the recovered signal will be stressed. Several types of test signals that have specific applications to test for signal impairment have been developed. These signals are either transmitted continuously or inserted on the top lines in a television frame.

Normal programming has to be removed from service if testing is to take place continuously for every line in a frame. Normally the signals appear inserted on the top lines of a field, often lines 17 and 18. Line 17 may contain the composite test signal and 18 may contain the combination test signal. These signals are shown in Figs. 4.13 and 4.14. Use of these signals and the test methods were recommended in NTC-7 (Network Transmission Committee No 7). Lines 14 through 21 appear at the top of the television screen, so they are essentially not seen unless the picture height control is reduced. These calibrated test signals are often referred to as VITS (vertical interval test signals). The amplitude of the standard video signal that is the output from the television cameras, video production switching boards, and distribution amplifiers is 1 V peak to peak, which corresponds to peak white level and the negative horizontal synchronizing pulse. A calibrated oscilloscope with a bandwidth of 10 to 15 MHz, a calibrated vertical range that includes 1 V, an adequate triggering circuit, and a television horizontal time base selector can be used. A special type of oscilloscope with all the features needed to test the video waveform signal is called a *waveform monitor*. The higher-quality waveform monitors have a line select feature that permits selection of the vertical interval's top 21 lines, thus having the ability to select the desired vertical interval test signals. The amplitude scale is divided into IRE units (Institute of Radio Engineers, a forerunner of the IEEE). The 1-V peak-to-peak standard video signal corresponds to 140 IRE units marked on the graticule of the waveform monitor. Therefore, 100 IRE units, which is the range from white to blanking level, corresponds to 0.714 V. An example of a waveform monitor graticule is shown in Fig. 4.15. Different parts of the test signals are used to test video performance, and their departure from the normal waveform gives a measure of a particular type of video signal distortion causing picture impairment.

Since proper color television operation depends on the relative phase changes between the color burst signal and the chroma carrier, measurements of phase changes have to be made accurately. These measurements are made with a special instrument called a *vector scope*, which like the waveform monitor is a specialized oscilloscope. The basic screen trace is not horizontal like the waveform monitor or oscilloscope but circular. The departure from the circular trace corresponds to phase vectors. The screen graticule for a vector scope is shown in Fig. 4.16 with a typical trace showing the standard color bars. The test configuration using both the waveform monitor and the vec-

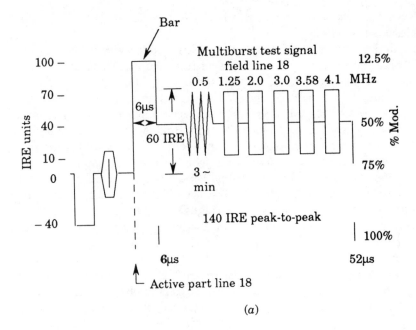

(a)

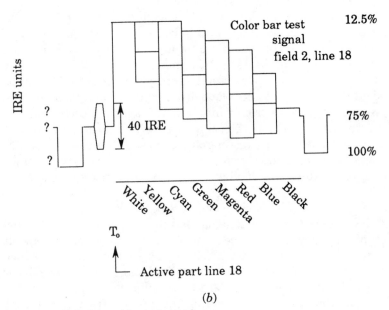

(b)

Figure 4.13 Television video test signals.

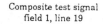

Composite test signal
field 1, line 19

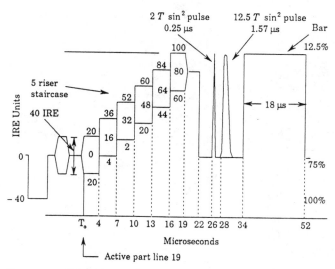

Figure 4.14   Television video test signal composite.

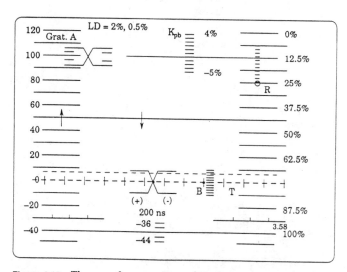

Figure 4.15   The waveform monitor graticule.

tor scope is shown in Fig. 4.17. To test a television transmitter, a precision demodulator is needed to detect or demodulate the television signal carrier and provide the base band television signal to the input of the waveform monitor–vector scope combination. Essentially the waveform monitor and the vector scope operate in the same fashion as

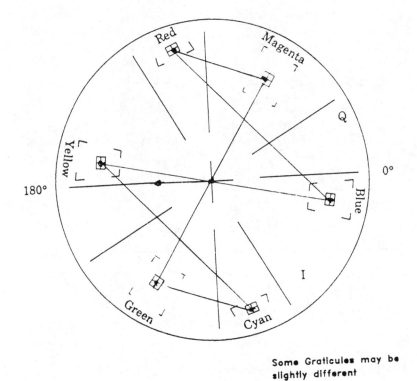

Some Graticules may be
slightly different

Figure 4.16 Vectorscope screen.

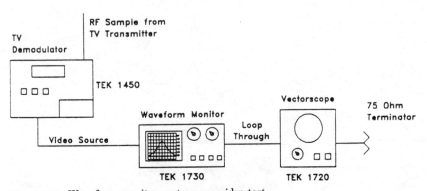

Figure 4.17 Waveform monitor-vectorscope video test.

the oscilloscope, that is, it is a high-impedance device. The 75-ohm terminating resistor, usually built into a BNC connector, has to end the test chain. This is necessary so that the instruments can measure the precise voltage level across the 75-ohm terminating resistor. If the test setup is connected across a 75-ohm terminated device and the loop through has a 75-ohm terminating resistor, then a double termination results, causing the voltage to be reduced by a third. This decrease of the expected voltage of 1 V by a third should alert the operator of a possible double termination.

4.2.2.2 Testing of NTSC signal quality. Since the United States uses the NTSC video waveform for the standard broadcast color television industry and NCT-7 is still generally the accepted standard, emphasis will be made in this book to such testing. However, the European standards of SECAM (sequential color with memory) and PAL (phase alternation line rate) are similar in many respects and some instruments have the usually optional capability of measurements in all three standards. The NCT-7 testing will be taken up on a test-by-test basis with reference to the test signals that are originated by a precision video test generator such as the TEK TSG-100, TSG-130A, 170D. Since full field (test signals on every active picture line) means that normal television service has to be suspended. The usual method is to insert these test signals in the vertical interval so that such testing does not interrupt normal television service.

Test 1, insertion gain (NCT-7, Sec. 3.2). Insertion gain is a test to measure the peak-peak amplitude of the video signal. The FCC composite test signal is used for this test, as shown in Fig. 4.14. Errors in the insertion gain affect the picture brightness as well as the color saturation. The long bar top should be at the 100 IRE level on the waveform monitor and the sync tip at the bottom of the trace at −40 IRE. This means that the test peak-to-peak level is at 140 IRE units, which corresponds to 1-V peak to peak. The blanking should be at zero IRE units. The correct condition is shown in Fig. 4.18. If the waveform monitor has cursors at top and bottom of the waveform monitor graticule (TEK 1780R), these can be set to 100 and −40 IRE units and the signal expanded between the cursors for a simple quick measurement.

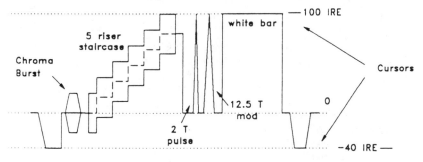

Figure 4.18 Proper insertion gain.

Test 2, line time distortion (NCT-7, Sec. 3.4).   The same FCC composite test signal is used for this test and the flatness of the long white bar is the parameter to be examined. This results in what is known as *line time distortion*. If the tilt of the white bar top is poor, the picture produced with this type of distortion will result with the brightness varying across the picture from left to right. Also, some horizontal streaking and smearing will reduce the detail of scenes with sharp vertical edges. This type of distortion is shown in Fig. 4.19.

Test 3, short time distortion (NCT-7, Sec 3.5).   Testing for short time distortion is performed on the rise time edge of the long white bar portion of the FCC composite signal. This type of distortion also affects vertical edge detail in the television picture. Another technique used tests for such linear distortions by using the K factor rating, which measures ratios of the 2T pulse to the white bar levels. The special area of the waveform monitor (K_{pb} scale) is used. The TEK 1780R has a special K factor scale marked on its graticule which can be used for weighting the measurements. However, the NCT-7 method will perform adequately for this type of distortion where errors on the rise

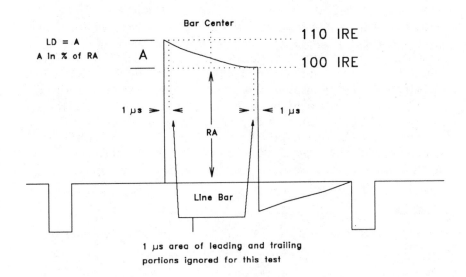

Some test simply by measuring the
White Bar overshoot as for example 10 IRE.
NTC−7 test measurement is made
in %, for this case 10%.

Figure 4.19   Test 2 line time distortion example.

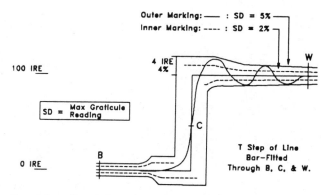

Figure 4.20   Test 3 short time distortion.

time portion of the white bar are given in percent of maximum graticule reading. This type of distortion is shown in Fig. 4.20.

Test 4, field time distortion (NTC-7, Sec. 3.3).   This test measures the level of the video signal over a section of the whole field and essentially has to be performed with the television service removed. The test signal has to be made with either a full-field square wave or a window signal. The window signal, when viewed on a television monitor, will show a black screen with a white square approximately in the center. The video waveform has 130 lines in the center of the field with each of the 130 lines containing 18-microsecond (μs) white pulses ($1 - T$ step), thus creating the window. The test, using the full-field square wave, is made by measuring the peak-to-peak level deviation of the bar tilt, as shown in Fig. 4.21. Distortions caused by 60 Hz power line hum are considered as field time distortions, since the power line frequency is about twice the field rate.

Test 5, chrominance-to-luminance gain inequality (NTC-7, Sec. 3.6).   This test is a measurement comparing the gain of the chrominance signal components and the luminance or brightness components. Such errors in these signal gains cause color saturation problems, that is, the reds redder or faded, etc. Chrominance-to-luminance gain and delay are also made on the FCC composite test signal or the similar NTC-7 composite test signal by making measurements on the 12.5T pulse. This signal has the shape of a sine-squared pulse with 3.58-MHz modulation. The half-amplitude duration (HAD) is 1.56 μs with a full-pulse duration of 3.12 μs. The correct signal with no gain or delay problems is shown in Fig. 4.22. The T duration is defined as the Nyquist interval of $1/f_c$ where f_c is at 4 MHz, the video testing bandwidth, which corresponds to 125 ns. Essentially the baseline portion of the 12.5T modulated pulse is to be examined.

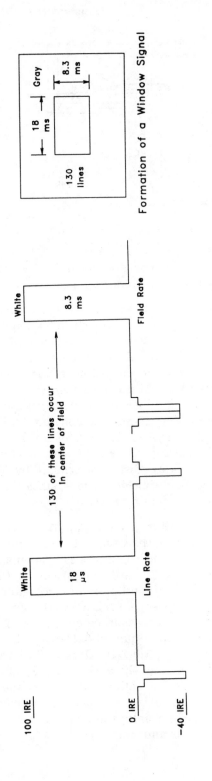

Figure 4.21 (a) Formation of a window display.

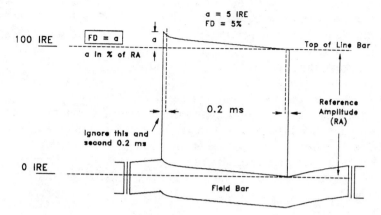

Figure 4.21   *(Continued)* *(b)* Measurement of field time distortion.

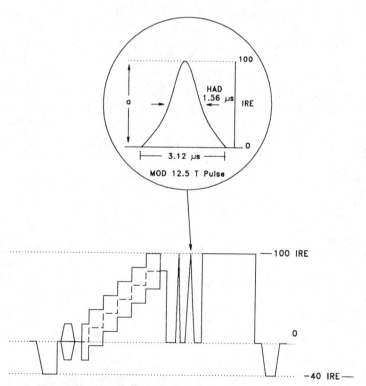

Figure 4.22   The modulate sine-squared pulse.

Test 6, chrominance-to-luminance delay inequalities (NTC-7, Sec. 3.7). This test is conducted on the same portion of the $12.5T$ modulated pulse as the chrominance-to-luminance gain inequalities. The chroma delay distortion causes what is known as color bleeding or smearing at sharp edges of an object. Errors in chrominance-to-luminance gain and delay for various cases are shown in Fig. 4.23. A chart or graph is used to complete the measurement if an ordinary waveform monitor is used, such as the TEK1480 series of instruments. More advanced instruments, which contain the essence of the nomograph in their computation circuitry, are able to make the measurement directly with the value printed on the screen. The TEK1780 or VM-700 are instruments that will perform this measurement directly. This nomograph is shown in Fig. 4.24.

Test 7, gain-frequency response (NTC-7, Sec. 3.8). Testing the frequency response, as most of us know, is a measure of how a system passes various frequencies without altering the amplitude. The FCC combination test signal with six equal amplitude bursts of selected frequencies is used. This signal appeared previously in Fig. 4.13a). Frequency response problems cause many types of video picture impairment in both color and brightness as well as picture sharpness, that is, resolution. The multiburst is a burst of equal-amplitude frequencies that fall in the video bandwidth and are 0.5, 1.25, 2.0, 3.0, 3.58 and 4.1 MHz. Measurement of the peak-to-peak amplitudes of the packet of the highest frequency are made either with reference to the white bar or the lowest-frequency amplitude. Since this is a relative measurement, the difference each system should settle on is whether the white bar amplitude or the amplitude of the first (0.5 MHz) packet is the refer-

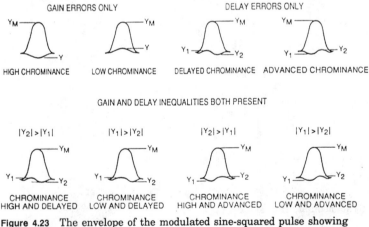

Figure 4.23 The envelope of the modulated sine-squared pulse showing error in chrominance-to-luminance gain and advance delay inequalities.

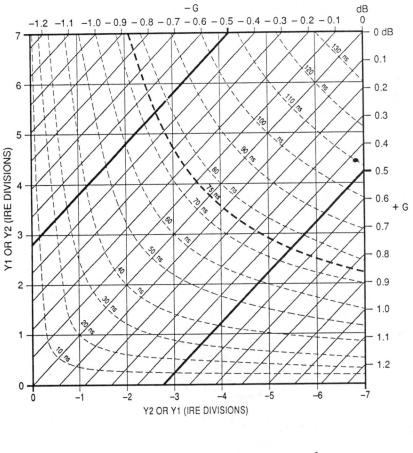

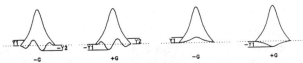

Figure 4.24 Measurement convention and nomograph test 5 and 6 chrominance-to-luminance gain and delay inequalities.

ence. An example of frequency-response high-frequency roll-off is shown in Fig. 4.25. Use of cursors available on more advanced waveform monitors such as the TEK1780 simplify the measurement.

Test 8, differential gain (NTC-7, Sec. 3.11). This measurement is made on the portion of the FCC composite test signal or the NTC-7 composite test signal that is known as the five-riser staircase. This portion consists of constant-amplitude chroma frequency superimposed on various (six) levels of luminance. Severe cases of differential gain can be

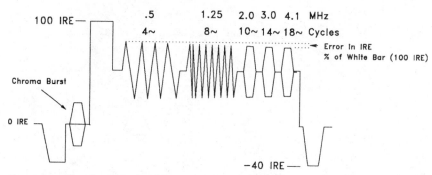

Figure 4.25   Test 7—example of frequency response errors.

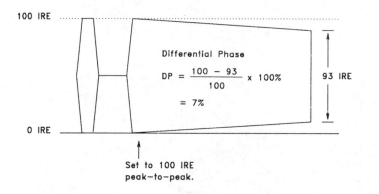

Staircase of Chroma passed through a Chrominance Filter
normally built into the Waveform Monitor

Figure 4.26   Test 8 differential gain.

seen and measured on a vector scope. However, it is more common to use the waveform monitor by passing the signal through the built-in chrominance filter. This removes the luminance portion and shows on the screen the chroma-level variations that occur at the various luminance levels more clearly. The five-riser staircase is magnified over the screen of the waveform monitor and passed through the filter to give a display as shown in Fig. 4.26. This test is made by measuring the difference between the peak-to-peak amplitude of the lowest burst level and the highest burst level expressed as a percent. The TEK1780 series waveform test equipment simplifies the measurement as well as provides a printed screen value. The TEK VM-700 equipment makes this measurement accurately with the touch of a button.

Test 9, differential phase (NTC-7, Sec. 3.14).   Again the test of differential phase is made on the modulated five-riser staircase signal. Essentially

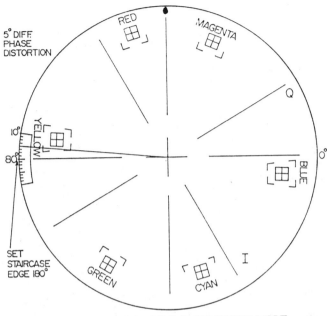

SET VECTOR TO CIRCLE IN VECTOR MODE

Figure 4.27 Test 9 differential phase.

the chroma phase changes that occur at the five luminance levels are examined. Usually the peak-to-peak phase changes are recorded as differential phase. Improper color and shifts in hue are caused by this type of distortion. To make accurate measurement of small changes in differential phase, a vector scope with a differential phase feature is needed. The TEK VM-700 makes an automatic measurement of this parameter. A vector display of a five-degree error in differential phase is shown in Fig. 4.27. Use of a single trace on the TEK 1780 along with the vector scope can suffice, but the double-trace method of the TEK 1780 can provide more accurate results. Again, the TEK VM-700 automates this measurement parameter.

4.2.2.3 Video noise testing. Noise in video systems basically causes a grainy picture on a busy background. Video noise is seen on a waveform monitor as baseline grass on the horizontal portions of the video test signals. To analyze noise, a series of weighting filters is needed in the signal path of the waveform monitor. The TEK 1480 can be field-fitted with noise-measuring capability. The TEK VM-700 has built-in filters (digital) and computational features that provide automatic noise measurements.

Use of the TEK 1430 random noise measurement set allows the test

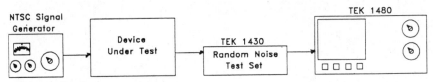

Figure 4.28   Video signal-to-noise test.

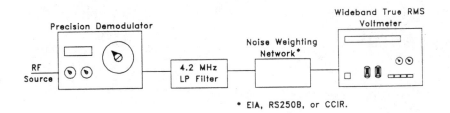

* EIA, RS250B, or CCIR.

Procedure: (1) Full modulation, make reading on voltmeter
            (2) Remove modulation signal. Read noise on meter.
            (3) If meter has dB scale, difference between signal
                level and noise level is S/N ratio.
            (4) If meter does not have dB scale, calculate dB S/N.

Limiting factor is the residual noise from the Demodulator.

Figure 4.29   Video S/N measurement.

signal (VITS) to pass through to a waveform monitor, and a noise pedestal is inserted. Matching of the signal noise to the inserted noise on the pedestal allows the measurement to be made. This is a comparative technique that provides accurate results. This method is shown in Fig. 4.28. The video base-band signal-to-noise ratio can be tested on an RF carrier (that is, a transmitter test) by using a precision demodulator, video filter, and weighting network and an rms voltmeter (wide-band) or oscilloscope. Such a test setup is shown in Fig. 4.29. The principal limiting factor in this test procedure is the quality of the demodulator. Some types of low-cost demodulators have a base signal-to-noise ratio of 55 dB, making this is as low a signal-to-noise ratio as can be measured.

4.2.2.4 VSB AM video modulation parameters. The AM type of video modulation for NTSC transmitters uses negative modulation, where the sync tip is at full RF power. The zero carrier level point corresponds to 120 IRE, where peak white level corresponds to 100 IRE units. The overall peak-to-peak percent of modulation is usually at 87½ percent. Most commercial television transmitters have built-in monitoring meters and instruments coupled to alarm circuits which warn the operator in charge of any problem with transmitter performance or out-of-tolerance conditions. The aural transmitter portion of the television transmitter may have a power output between 10 and 20 percent of the total peak power output. The audio signal frequency modulates a 4.5-MHz subcarrier which appears at 4.5 MHz ± 1 kHz above the video carrier frequency. This carrier is frequency-modulated, where 100 percent modulation corresponds to ±25 kHz. Audio bandwidth is specified at 30 Hz to 15,000 Hz using the standard preemphasis curve of 75 μs.

Video modulation testing. Video modulation can be tested using an RF-type spectrum analyzer such as the TEK models 7L12, 2710, 490 instruments. Essentially the instrument is operated in the zero-scan linear mode, which acts like a sensitive calibrated receiver, where the bottom graticule on the screen corresponds to zero carrier voltage. The analyzer is connected to an RF test port and the center frequency is set at the test carrier frequency with a resolution of 3 MHz. Using the fine-tuning peak, set the signal for maximum upward deflection and adjust the IF gain control to place the video sync tips on the top graticule. Then switch to the linear mode and observe that the picture white level is at the graticule line one up from the bottom. Therefore seven squares corresponds to 87½ percent modulation over 100 IRE units of video signal, meaning each square is at 12.5 percent per square. This method is illustrated in Fig. 4.30. The video signal can be utilized for approximate results. However, an NTSC signal generator with a flat field selected as the signal source used to drive the transmitter, will make this measurement easier and the results more accurate.

Television audio transmitter tests. The aural transmitter section can be tested for percentage of modulation using an accurate deviation meter with a test signal generator. FM deviation meters are commercially

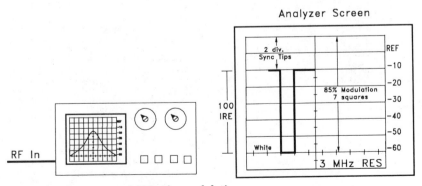

Figure 4.30 Percentage of AM video modulation.

available from several manufacturers. However, a good spectrum analyzer with a signal generator can also accurately measure the FM deviation by the Bessel null method. It must be remembered that the carrier frequency is deviated by the amplitude of the audio signal. The measurement of the frequency deviation by the Bessel null method will be taken up in the section covering FM stereo radio service in Sec. 4.2.3. The overall amplitude of the video and audio carrier level can be measured using a spectrum analyzer by connecting to an RF port for a sample of the TV transmitter's output. This can be done with the transmitter on line. The test setup and a screen display of the television transmitter output are shown in Fig. 4.31. This test indicates the signal-level differences between the video and audio carriers as well as the bandwidth and the carrier-to-noise level. These are all very important measurements and are easily and accurately made using a high-quality spectrum analyzer.

Many commercial television broadcast stations presently transmit the audio portion in the BTSC (Broadcast Television Stereo Channel) stereo format. The left plus right (L&R) channel constitutes the monaural audio signal, which frequency-modulates the 4.5-MHz standard television audio subcarrier. The left minus right (L-R) signal frequency-modulates a second subcarrier operating at twice the horizontal sync frequency. Also, a second audio program (SAP) subcarrier operating at five times the horizontal sync provides a second language audio channel. Another subcarrier at 6.5 times the horizontal sync provides for professional channel use. This signal layout is shown in Fig. 4.32. Since this signal is constantly varying with modulating program material, it is very difficult to make accurate measurements with a spectrum analyzer. A better method is to use one of the com-

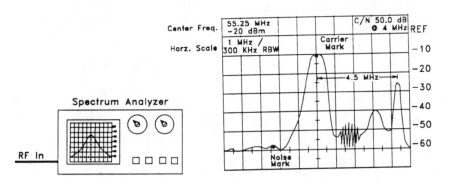

RBW = Resolution Bandwidth
C/N measured between Carrier level
    and noise level at markers.

Figure 4.31 Television RF carrier measurements.

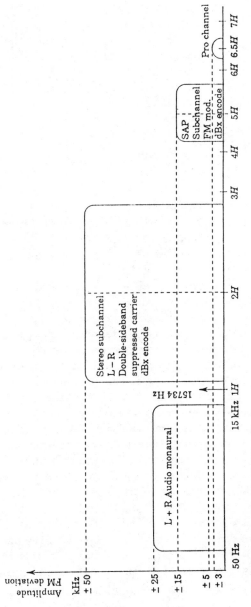

Figure 4.32 Basedband BTSC audio format.

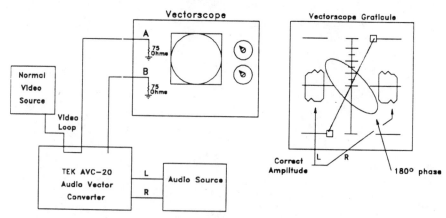

Figure 4.33 Use of TEX AVC-20/Vectroscope for BTSC audio testing.

mercially made instruments to test for subcarrier deviation and overall stereo separation. One instrument manufactured by FM Systems for cable television measures the BTSC stereo performance with an accuracy of 1 percent. Measurements of pilot deviation and stereo separation are possible using this instrument at either the standard television IF frequency of 41.25/45.75 MHz or 4.5 MHz TV stereo modulated signal. The instrument can be tuned to a television converter channel 3 for stereo separation on any television carrier. For television station use or a studio facility providing television satellite service, Tektronix has several instruments capable of measuring and monitoring television stereo audio. The simplest is using a vector scope in combination with the TEK AVC-20 Audio Vector Converter. An electronic graticule is provided by the AVC-20 so that the vector scope can be used with its normal graticule for ordinary video monitoring. Activation of the AVC-20 then allows for correct amplitude and phase measurement on the audio signal. Connections and a graticule layout for a correct audio signal are shown in Fig. 4.33. The TEK 75BTSC Aural Modulation Monitor Decoder is a separate instrument used to monitor the stereo audio performance. Bar graph readouts display all the required signal components with sufficient accuracy to ensure proper performance. The TEK 760 Stereo Audio Monitor displays phase measurements like a vector scope with the addition of bar graph displays of left and right signal amplitude. The selection of instrumentation depends on whether the requirement is ongoing monitoring or for occasional measurements.

Video and audio carrier frequency tests.   The television video and audio carrier frequencies are required by the FCC to be within specified lim-

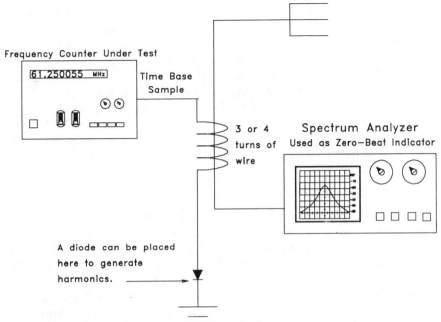

Figure 4.34   Frequency check using WWV standard.

its at all times. Measurements have to be periodically made and the results noted in the station's logbook. Cable television operators also are required by the FCC to maintain frequency accuracy with less-tight tolerances for the signal-generating equipment in their hub or head-end sites. Television broadcast stations maintain a carrier frequency tolerance of ±1 kHz for the video carrier and the 4.5-MHz audio subcarrier as well. Most stations prefer to stay within the ±200-Hz limit. The most preferred frequency standard is the 60-kHz standard transmitted by the National Bureau of Standard station WWVB. This standard can be used to check the calibration of a stable frequency counter. The counter can be connected through a directional coupler to the transmitter RF output to obtain an RF sample for frequency measurement or it may obtain a sample of the carrier excitation equipment output. When using a frequency counter, the modulation has to be removed for proper counter operation. Most transmitters have a frequency monitor built in that compares the exciter generator frequency with a standard frequency and displays the difference as a frequency error. If a frequency counter is connected to the exciter before the modulating section, it too can be used to monitor frequency accuracy. A method of testing a frequency counter for accuracy against the

WWV standard of 10 MHz is shown in Fig. 4.34. This method inserts the frequency counter's time base signal together with the WWV carrier. The beat difference as observed on the spectrum analyzer corresponds to the error. A communications receiver can be substituted for the spectrum analyzer and earphones can detect the difference. For instance, if the counter's time base is 10 MHz and WWV carrier at 10 MHz is tuned in by the receiver or analyzer, zero beat will occur if there is no frequency difference. The frequency of the tone is the frequency difference between the counter time base and the WWV carrier. A small signal diode may need to be inserted in series with the coil loop to generate time-base harmonics.

4.2.3   FM stereo broadcast service

FM stereo broadcast service is a much-used and desired service by the American public. Home stereo receivers and automobile stereos are owned by most families and often as much or more money is spent on this form of audio entertainment as home video equipment. The FM stereo and mono broadcast stations occupy the frequency spectrum space of 88 to 108 MHz, which contain 100 FM channels spaced 200 kHz apart. FM channels adjacent to one another are not allowed to operate in the same geographical area so as to prevent any cross-channel interference.

4.2.3.1   FM stereo signal parameters. For stereo operation, the left channel is summed with the right channel (L&R), thus forming the monaural signal (mono), and the left-minus-right channel amplitude modulates a suppressed 38-kHz carrier. A 19-kHz pilot carrier is also included and the whole L&R, L−R (AM), 19-kHz pilot, signal FM modulates the RF carrier. A spectrum for this multiplexed stereo signal is shown in Fig. 4.35. The 19-kHz pilot carrier with an accuracy of ±2 Hz is used by the receiver to generate by doubling the 38-kHz carrier needed to demodulate the L−R signal. The L&R and L−R signal are added and also subtracted to produce a 2R and a 2L signal which when level-adjusted produces a proper right and left channel. The 38-kHz carrier should be suppressed to a level sufficient to limit its frequency modulation of the main carrier to 1 percent. A common method used to generate the stereo signal employs a switching method which switches between the left and right channel signals at a 38-kHz rate, which essentially generates all the signal components (L&R, L−R) necessary for stereo FM transmission. This switching technique also generates harmonics which are filtered using a lowpass filter. Some of the switching noise may appear in receivers, particularly if the filter does not provide adequate attenuation in the stop band or is

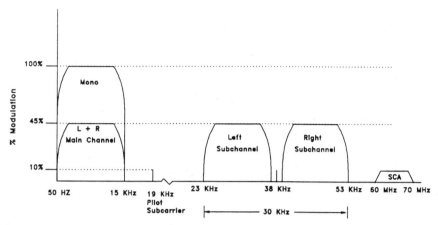

Figure 4.35   FM stereo signal.

not sharp enough at the cutoff frequency. This switching noise appears in a stereo receiver as hiss. A simple test of stereo hiss is to switch the receiver to the mono position and note the amount of hiss reduction. The FM preemphasis/deemphasis is an R-C network with a 75-μs time constant. This is commonly used in FM stereo broadcast stations, except those that use the Dolby system.

4.2.3.2 FM stereo broadcast tests.   Tests on FM stereo systems fall into two areas. The first area concerns the stereo signal as it goes through the switching and conditioning circuits to the stereo generator that in turn feeds the transmitter. The second area concerns the transmitter, where the stereo-generated signal is transmitted over the air. Tests on the signal-to-noise ratio of the left and right channels as well as the frequency response and stereo separation have already been covered in Chapter 3. Testing on proper modulation of the transmission equipment is what will be taken up now. The basic FM stereo broadcast system is diagrammed in Fig. 4.36. Signal to the stereo generator is the separate left and right channels. Therefore, standard audio response, distortion, and stereo separation tests are used. Through the stereo generator a spectrum analyzer can be used to examine the BTSC stereo signals, usually using single tones for the left and right channels.

Proper FM modulation of the transmitter is extremely important to the quality of the FM stereo signal. It is suggested that a review of FM fundamentals be taken up at this point. In brief, the carrier frequency is deviated up or down at an audio rate. For FM broadcast the deviation is ±75 kHz for a total deviation of 150 kHz. For the monaural case, this amount of deviation corresponds to 100 percent modulation. For the stereo multiplex case, 100 percent modulation corresponds to

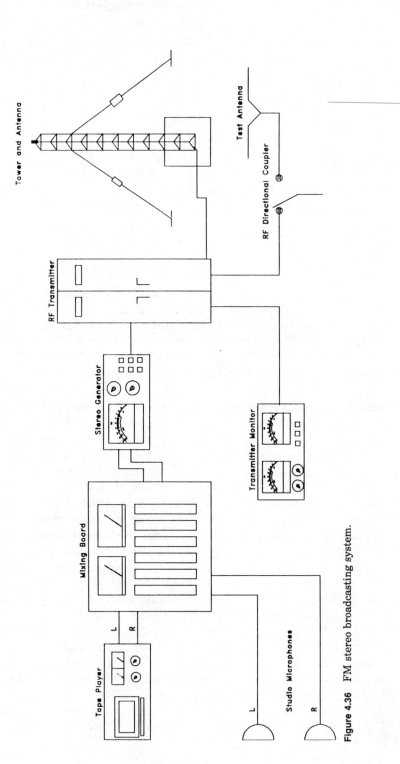

Figure 4.36 FM stereo broadcasting system.

only 45 percent of the total deviation of 67.5 kHz (±37.5 kHz). A high-frequency spectrum analyzer (TEK2710) can be used to test the transmitter deviation at radio frequency by taking an RF sample of the transmitter output through a directional coupler or sample loop antenna. Using the calibrated span on the spectrum analyzer and a pure audio tone (1000 Hz) as the modulating source, a measure of deviation can be made. The connections and spectrum analyzer display are shown in Fig. 4.37.

Another method of testing for the proper FM modulation is called the Bessel null method. In the modulating process the carrier is deviated plus and minus and an infinite number of sidebands are generated. The amplitudes of these sidebands vary with the index of modulation, which is defined as the peak carrier deviation divided by the modulating highest frequency. The amplitudes of the sidebands vary according to a series of Bessel functions, giving the method its name. At certain indexes of modulation, the carrier essentially disappears, causing a null. At this point all the energy appears in the sidebands. Mathematical analysis of the frequency modulation pro-

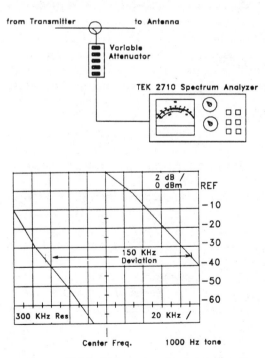

Figure 4.37 FM deviation test using a spectrum analyzer.

TABLE 4.2

Null	Modulation index
1	2.405
2	5.520
3	8.653
4	11.792
5	14.931
6	18.071
7	21.212
8	24.353

cess results in the table of nulls and modulation index as given in Table 4.2.

Since the amplitude of the modulating signal determines the carrier frequency deviation, it in turn affects the modulation index. Also, the modulating frequency is a determining factor of the modulation index. Therefore, the procedure in the Bessel null method for determining the modulation percentage is to calculate a modulating frequency that will produce a carrier null. For example, at 100 percent modulation (a peak deviation of ± 75 kHz) a modulation index for a first null will be $f = 75$ kHz/2.405 = 31.185 kHz. This frequency is out of the audio pass-band, so the second null will be tested, $f = 75$ kHz/5.520 = 13.587 kHz which falls in the higher frequency of the audio range, and which should work. The equipment should be connected as shown in Fig. 4.38. A frequency counter is necessary to adjust the modulating audio frequency to the required precision. First the audio frequency required for a second null at 100 percent modulation (13.587 kHz) is selected. Now the amplitude from the audio generator is increased to increase the peak deviation. The first null as observed on the spectrum analyzer tuned to the carrier center frequency is not at 100 percent modulation. The second carrier null will be at 100 percent and should be as observed on the spectrum analyzer screen illustrated in Fig. 4.39. At 75 percent modulation the peak deviation will be 0.75×75 kHz = 56.25 kHz, which will give a modulating frequency for a second null of $f = 56.25$ kHz/5.520 = 10,190 Hz. Now a chart can be made so the Bessel null method can be used to check the calibration of the modulation monitor. This chart is shown in Table 4.3.

The modulation monitor is actually a precision demodulator (receiver) with a built-in deviation meter. The Bessel null method is a good procedure for testing the accuracy of the monitor. Both methods of deviation testing require essentially removing the FM service and substituting the precision audio generator. Once the modulation monitor has been tested, it can be used to observe the peak deviation and hence the day-to-day operation of the FM transmitter. For stereo op-

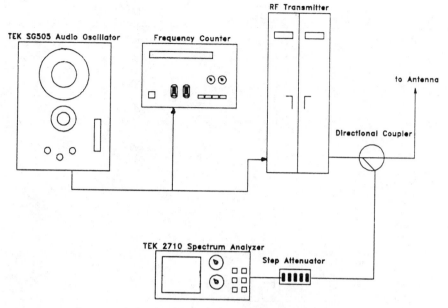

Figure 4.38   Connections for the Bessel null method of FM modulation percentage.

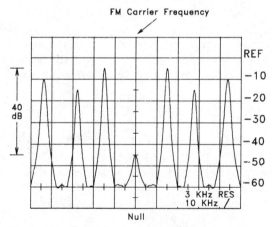

Figure 4.39   Example of screen showing second null.

eration, the 19-kHz pilot and the 67-kHz SCA carrier can be set accurately by using the Bessel principle. Since the modulation percentage is low, they can be considered as narrow-band FM carriers with sum and difference frequencies centered about the FM carrier. The 19-kHz pilot is required to be set at 10 percent modulation, which corresponds

TABLE 4.3 Modulating Frequencies Required to Produce Carrier Nulls for the Range of Modulation Percentages

% Modulation	Null	Modulation Frequency
100	2	13,587
75	2	10,190
50	2	6,793
45	2	6,114
40	2	5,435
30	1	9,355
10	1	3,125
8	1	2,495
1	1	312

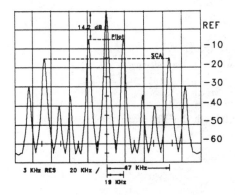

Figure 4.40 Example of stereo pilot and SCA carrier amplitudes as shown on spectrum analyzer screen.

to 14.2 dB below the carrier and for the 67.5 kHz SCA 25.0 dB below the carrier to produce a spectrum analyzer display as shown in Fig. 4.40.

The RF transmitter in many ways operates in a similar fashion to standard AM broadcast in AM radio as well as television service. The carrier frequency has to be tested accurately using some of the methods presented in the earlier chapters. Power output, antenna, and transmission line tests have to be performed on the transmitter system. The audio and television signal quality have to be tested going into the transmitter and out of the transmitter by use of precision demodulation equipment and the resulting signal quality tests. Problems with the transmitting system in altering or distorting the signal can then be identified, measured, and corrected. The test procedures and methods of proofing any broadcast station, either AM, FM, or television, are available from the FCC, and the specifications are contained in the pertinant rules and regulations. The National Association of Broadcasters publishes an engineering handbook which specializes in broadcast testing procedures and performance specifications.

Microwave Systems Testing and Measurements

Microwave systems appear in four major areas, communications, radar, medical, and consumer electronics.

The medical uses are highly specialized and consist mainly of RF heating equipment used to direct heat to various areas of the anatomy. The consumer microwave devices also are used for heating, principally in microwave ovens. This chapter will be devoted to the two largest areas of microwave devices, communications and radar. Communications can be further broken into two groups, line-of-site terrestrial microwave communications and satellite up-down communication links. Both of these systems are used for voice, video, or alphanumeric communication employing various modulation techniques.

In the past, the telephone companies provided video communications to the various television networks through the many microwave line-of-site links running from coast to coast. The television broadcast industry and the telephone companies agreed on the necessary video standards the microwave links needed to transmit high-quality television pictures. The published results, found in NTC-7 (Network Transmission Committee No. 7) specifications and signal standards are still in use today. Subsequently, many private companies have formed and obtained the necessary FCC licenses for operating terrestrial line-of-site microwave links. Some railroads obtained licenses and constructed microwave facilities along the railroad rights of way. Railroad control information was relayed back and forth along the bidirectional microwave paths. However, the railroads only needed a very small amount of the link's capacity, and so they leased channel space to the telephone companies as well as other users.

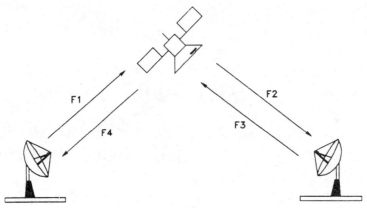

Figure 5.1   Satellite communication system.

Many microwave links are operated bidirectionally with communication lines flowing either way.

Satellite communication was actually an outgrowth of the space age. Once the techniques of constructing a satellite and placing it into orbit around the earth were developed, the need to communicate with the satellite became evident. Most communication satellites act as transponders, where many channels are transmitted to the satellite on one group of frequencies and retransmitted to earth on another group. This concept is shown in Fig. 5.1. Oftentimes there will be only one earth-based transmitting station and many earth-based receive-only stations. Such a system is used to provide programming to the many companies providing cable television service in the country. Satellite systems providing data communications and telephone services are operated continually in the bidirectional mode.

Radar systems operate on the principle of transmitting a narrow, high-power pulse of energy with a long duty cycle. This pulse of energy is transmitted in a narrow beam that strikes a metallic target. Some of the reflected energy is received at the transmitting site during the pulse's off time. The time between the transmitted and receive pulse is a function of the distance between the transmitter and the target. This time is computed into a distance between transmitter and target. A large parabolic antenna sends a narrow beam toward the target, and the angles at which the antenna points are interpreted as azimuth and elevation angles to the target. An early use of radar was in war time for artillery gun control. Modern-day radar systems are operated as early warning defense radars, air traffic control radars, marine navigation and air navigation radars, as well as the usual weapons control radars.

5.1   Line of Microwave Communications

Microwave communication systems operate in the line-of-site mode simply because microwave energy is subject to refraction by the ionosphere. Microwave energy in the gigahertz frequency range is propagated through space in a similar fashion to light waves. The microwave transmitter generates the high-power carrier, which in turn is coupled through a feedhorn antenna to a parabolic reflector. This antenna assembly forms a narrow beam which is focused toward the receiving antenna assembly. The receiving antenna is the same type as the transmitting antenna and concentrates the received energy to the feed horn, which is coupled to the receiving electronics. Generating microwave energy at high power levels is necessary for obtaining good communication distances. High receiver sensitivity is also needed to increase the distance between the transmitter and receiver.

5.1.1   The reflex klyston

The reflex klyston vacuum tube is still used in today's microwave facilities as a generator of microwave energy. This device uses a resonant cavity as the frequency-determining element. The multicavity klystron acts as a power amplifier at microwave frequencies. A tuning strut system that changes the shape of the resonant cavity allows the reflex kystron to be tuned through a narrow band of microwave frequencies. The electrode voltages, principally the repeller voltage for the reflex klystron, is critical to sustaining oscillations. Varying the repeller voltage slightly allows the frequency to be changed. Hence the reflex klystron is able to be frequency-modulated by modulating the repeller voltage. Reflex klystrons are used as the local oscillator in mixer circuits used in microwave transmitters, receivers, and signal generators. The multicavity klystron is used as a power amplifier in television transmitters, radar transmitters, and tropospheric scatter communication systems. Power in the range of 10 to 50 kW can be obtained by this microwave vacuum tube. If 50 kW exits an antenna system with 12 dB of gain, the resulting effective radiated power (ERP) is 800 kW.

5.1.2   The magnetron

The magnetron is another type of resonant-cavity vacuum tube with the capability of generating high-microwave power output. This device was developed to produce the high-power microwave pulses required by radar systems. The magnetron has several cavities placed around the central cavity containing the cathode. A strong permanent magnet structure is needed to sustain the oscillating currents in the cavities. The cathode is pulsed negatively 10 to 20 kV, causing a high-power microwave pulse of energy to be picked up by a loop in one of the peripheral cavities. This loop is coupled to a quarter-wave antenna via a short coaxial feeder line. The quarter-wave antenna, placed inside a waveguide, radiates RF energy into the waveguide, which in turn feeds the microwave energy to a load. The magnetron has a fairly low average output power. However, when this is concentrated into a narrow pulse, the peak pulse output can be as high as 20,000 W. The frequency of the magnetron is determined by the size of the peripheral cavities and by slight variations in the voltage. If the voltage is changed much, the magnetron may drop out of oscillation or the tube may be damaged. An elaborate mechanical system of moving metal slugs that change the cavity size will allow larger frequency changes. The largest cavity size determines the lowest output frequency of the magnetron.

5.1.3   The traveling wave tube

The traveling wave tube (TWT) is another device that acts as a microwave amplifier. This device, unlike the klystron or magnetron, can operate over a relatively large bandwidth. The tube consists of a cathode and beam-forming plates, a helix electrode, and an anode, all placed inside a glass-evacuated tube. An outside helix near the cathode couples the low-power microwave energy to the system and another outside helix near the anode extracts the large (amplified) output to the load circuit, usually through a coaxial feeder line or energized wave guide. No resonant cavity is used. However, a permanent magnet structure is needed to maintain the electron beam through the center of the interior helix electrode. Varying the interior helix voltage tunes the tube for best gain. Gains of 30 to 60 dB are attainable with power outputs to better than 10 W.

A variation on the TWT, the backward wave oscillator tube, can be tuned over an octave range by varying the helix voltage. However, by varying this voltage the signal amplitude changes as well. Therefore, special circuitry is required to correct signal amplitude with change in frequency by varying helix voltage.

By now it should be evident that the electrode voltages on all of these microwave-generating or -amplifying devices are extremely critical for proper operating conditions. Therefore, accurate measurements should be made periodically to maintain correct performance. Usually a logbook of system measurements is kept and these records will oftentimes provide a warning that a device is failing.

5.1.4   Solid-state microwave devices

Solid-state devices used in microwave equipment are Gunn-effect devices (diode) or IMPATT devices. When the bias voltages are adjusted properly, these devices can operate in the negative-resistance portion of their operating characteristic curves and thus sustain oscillations at microwave frequencies. Usually these devices are installed inside a waveguide structure, allowing the microwave energy to be efficiently coupled to the output. These devices can generate signals of only a few gigahertz and at low power of about 20 mW. This device can be coupled to a multicavity klystron power amplifier if greater power output is desired.

The GaAsFET (gallium arsenide field-effect transistor) amplifier can provide an output of about 2 W (33 dBm) in the 13-GHz band of frequencies. Most microwave amplifiers need some form of harmonic filtering on the output to eliminate spurious radiating frequencies.

5.1.5   Microwave instruments

Microwave instruments are connected to various points in an operating system by special coaxial or wave-guide devices. Most microwave communication systems provide such test points placed at various points in the system where important operating parameters should be monitored. The critical dc electrode voltages and currents are often measured by built-in digital multimeters which are switch-selectable. Klystron current is a very important parameter to be recorded in the logbook. If a built-in metering circuit is not available, the operating manual should indicate the required test parameters and test ports. Most microwave communication facilities have directional coupler test ports placed in each klystron feed.

5.1.5.1   The microwave power meter.   A power meter with a microwave diode or thermistor sensor can be connected to an RF test point and the klystron RF power output can be determined. A directional coupler measurement using a power meter is shown in Fig. 5.2. Microwave power meters are available that can measure powers up to 50 GHz at levels from −70 to +44 dBm (100 pW to 25 W), depending on which sensors are used. Such power meters are available with either analog (D'Arsonval) or digital meters. Typical accuracy of digital power meters are ±0.5 percent or ±0.02 dB on the dBm scale. Analog meters are essentially limited to the electromechanical meter movement accuracy of ±1 percent for a taut-band meter using a mirrored scale.

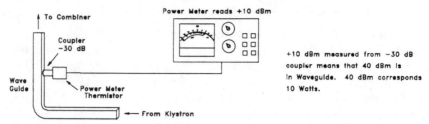

Figure 5.2 Microwave measurement of power at a directional coupler test point.

5.1.5.2 SWR measurement.

Impedance measurements of microwave RF components are usually made before a system is installed. However, in case of serious malfunctioning of the waveguide structure, the standing wave ratio (SWR) test will determine the faulty component. Usually moisture or water contamination with the usual corrosion causes the RF plumbing (waveguide or rigid coaxial cable) to become faulty. A microwave signal generator with a built-in 1-kHz amplitude modulation source, a slotted line, and an SWR meter can be used. Use of the AM 1-kHz source enables improved sensitivity for the SWR meter. It must be remembered that for the case of amplitude modulation, the amplitude of the detected signal is proportional to the RF carrier level. Hence the detection circuits in the SWR meter are sensitive 1-kHz circuits consisting of a square law detector decade (dB) attenuator and a 1-kHz amplifier detector. The setup for testing a microwave passive component, that is, a directional coupler, isolator circulator, or waveguide/antenna is shown in Fig. 5.3. This method measures the VSWR either as a ratio or in decibels. The impedance can be calculated using the well-known relationship $Z_1 = \text{VSWR} \times Z_0$, where Z_1 is the unkown impedance and Z_0 is the characteristic impedance.

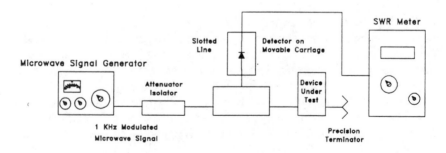

If Slotted Line Detector measures voltage maximum/minimum then SWR Is VSWR.

Figure 5.3 Testing impedance by VSWR.

5.1.5.3 Return loss bridge. A microwave return loss bridge coupled to a sweep generator can provide for a measurement of return loss in decibels as a function of frequency. This type of instrument is available in a complete package from some manufacturers or can consist of a separate microwave signal generator, return loss bridge, and a detector with an oscilloscope display. The oscilloscope has to have its sweep triggered at the start of the signal generator's microwave sweep. Also, the screen should be calibrated in decibels. The self-contained return loss bridge is a lot easier to use and accurate results can be obtained, particularly at microwave frequencies. However, precision RF bridges are available from several manufacturers with precision terminations in 50- and 75-ohm standard impedances. For microwave measurement in the millimeter wavelengths, waveguide bridges, terminations, and circulator/isolators are required.

5.1.5.4 Microwave frequency testing. Frequency measurements at microwave levels are performed using a high-quality microwave frequency counter. Frequencies up to 110 GHz can be measured using a special 11 digit display frequency counter with accuracies of 1 Hz for a continuous wave (CW) signal. Microwave frequency counters have extremely stable time bases and employ prescaling of the input signal. Naturally such instruments are expensive, and if only ballpark measurements are needed, possibly one of the tunable cavity-type wave meters can suffice. It should be obvious that accuracy and resolution overkill can ruin a laboratory budget. The wave meter is a tunable cavity with a frequency scale on the tuning knob assembly. Depending on the frequency range, coaxial or waveguide connections are selected. This instrument is used with a detector or power meter to determine when the cavity is tuned. The frequency is then read on the scale. For frequencies to about 13 GHz, coaxial types are available, and for frequencies to 40 GHz waveguide types are available. Accuracies range from about 0.2 to 0.15 percent of the measured frequency. This type of instrument essentially passes through its full power-off resonance and absorbs power when tuned to the resonant frequency. Hence this device is known as an absorption frequency meter.

Of course, a high-quality microwave spectrum analyzer can also provide a measurement of frequency if high accuracy is not required. The spectrum analyzer also provides a lot more information about the microwave signal, such as harmonics, spurious emissions, bandwidth, and modulation to name a few. This instrument is nearly a must-have device for proper microwave facility maintenance.

5.1.5.5 The microwave network analyzer. The network analyzer is an extremely useful tool, particularly for manufacturers of microwave equipment. This instrument essentially provides an analysis of the network, which may consist of one or more components. Network analysis provides information on the active or passive behavior of a network. Such parameters as the transfer (input/output) function, network impedances with magnitude, and phase are some of the measurements accomplished by a network analyzer. Modern network analyzers usually contain a swept frequency source as well as oscilloscope (CRT) or $X - Y$ recorder displays. Complex impedances for several network analyzers are presented on a CRT display with a Smith chart graticule. Network analyzers are available for frequencies from about 5 Hz to 100 GHz. In addition to simple gain/phase measurements, insertion loss, VSWR, and power can also be measured with a network analyzer. These instruments are usually expensive, and some are even computer-controlled with the accompanying computational programs with computer screen displays. Both active and passive devices, transmission-line device directional couplers, filters, circulators, and isolators can be tested using a network analyzer. Essentially this device is an out-of-service piece of test equipment and tests network or network components on a piece-by-piece basis. Before investing in a network analyzer, a review of what devices are available and at what cost is in order. Most manufacturers provide application notes covering a variety of instruments and test procedures, and copies can be obtained by contacting the area sales or applications engineer or by direct mail to the factory.

5.2 Microwave communications modulation

Microwave communication systems may employ several types of modulation or modulating techniques, depending on the use. Telephone service may consist of many narrow-band voice channels digitized and transmitted from point to point as a PCM (pulse code modulation) signal that may frequency-modulate a microwave carrier. For video information, the video signal may amplitude-modulate an RF carrier at one of the normal television carriers in the standard NTSC-VSB television format. Several of these carriers may then be up-converted to a band of microwave frequencies and then transmitted to a microwave receiver. Such a system is used to transmit cable television service to multiple points for service distribution and is referred to as CARS (cable antenna relay service). Network transmission of video television signals often requires a dedicated bidirectional microwave channel for point-to-point two-way transmission. In many instances, a microwave link between two points consists of combining multiple

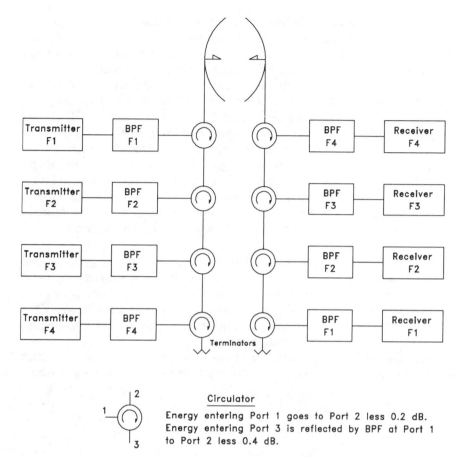

Figure 5.4   Example of a microwave system.

microwave transmitters contained in a band of frequencies and transmitting them to a matching bank of receivers. Such a system is shown in Fig. 5.4. The loss for transmitter 4 is greatest to the transmitting antenna (1.4 dB), but only 0.2 dB at the receiving antenna. Also, transmitter 1 only has 0.2 dB to the transmitting antenna and 1.4 dB from the receiving antenna to receiver 1. This is the same as for channel 4. Thus F1, F2, F3, and F4 are all received at the same level at the receivers. Typically, there are test points along the transmit/receive chain and in some instances a test probe may be required to provide the necessary instrument isolation.

5.2.1   The line-of-site microwave link

The line-of-site microwave link operating between the transmission site and the receiving site, which may be a distance of 10 to 30 miles, should be tested periodically and a log of such measurements and test results kept at each site. At the end of the construction phase of a microwave communication system or link segment, performance tests should be made to see if specifications are met or exceeded. The results should be submitted to the user as a certified written report. Representatives from the user or customer and representatives from the equipment manufacturer and construction contractor as well should jointly take part in the performance acceptance tests. The results of this test can act as a standard to be compared with other routine testing. Knowing the signal losses along the transmit/receive chain can be quite helpful in spotting any trouble spots. A fairly typical set of losses and gains along a single microwave hop is shown in Fig. 5.5. A fade margin of 40 dB or better is more than adequate to provide reliable microwave communications through various weather patterns.

5.2.1.1 Transmitter site tests. Measurements at the transmission site are always a good place to start. Usually the first sign of trouble is when the receive signal starts to decrease beyond the normal range. Some quite sophisticated microwave facilities have a low-power VHF radio telemetry system which can transmit such test parameters as transmitter power output, waveguide pressure, power supply voltages and currents, and building temperature. Most microwave facilities have pressurized waveguides with dried air or dry nitrogen gas which

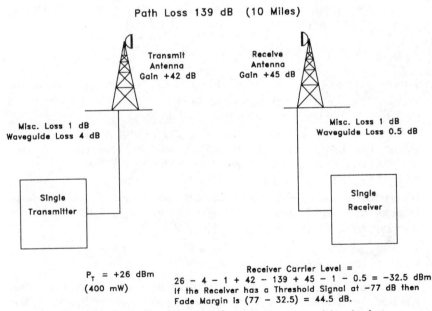

Path Loss 139 dB   (10 Miles)

Transmit Antenna Gain +42 dB

Receive Antenna Gain +45 dB

Misc. Loss 1 dB
Waveguide Loss 4 dB

Misc. Loss 1 dB
Waveguide Loss 0.5 dB

Single Transmitter

Single Receiver

P_T = +26 dBm
(400 mW)

Receiver Carrier Level =
26 − 4 − 1 + 42 − 139 + 45 − 1 − 0.5 = −32.5 dBm
If the Receiver has a Threshold Signal at −77 dB then
Fade Margin is (77 − 32.5) = 44.5 dB.

Figure 5.5   Example of gains/losses for a single microwave transmit/receive hop.

controls the waveguide loss and prevents moisture and corrosion. The klystrons as well as other electronic equipment are temperature-sensitive, and the transmitting and receiving site buildings must have proper heating or cooling equipment to maintain the temperature limits. Most of the power supplies at both the transmit and receive sites are metered, and the values of power supply voltage and current should be logged in a logbook. If the receive signal decreases beyond acceptable limits, the transmitter should be tested for proper power output, using an appropriate power meter, or with a microwave spectrum analyzer. A measurement of klystron current can also indicate that the klystron is not up to proper power output. The transmitter frequency should be tested because any frequency drift can cause the receive signal to decrease. It should be determined that the transmitter is on the prescribed frequency and at proper modulation and power output. The transmitter can be tested by testing these parameters at a test port, usually supplied on the transmitter unit by the manufacturer. A directional coupler is typically mounted on the transmitter unit or on the waveguide feed. A connector is provided for a power meter, the spectrum analyzer, or the frequency counter.

5.2.1.2 Receive site tests. At the receive site a spectrum analyzer placed at the waveguide feed to the receiver can indicate the level into the receiver. If this signal level is within limits, then possibly the receiver is out of the tuning range causing the problem. If this signal level is too low, then the level should be tested back toward the receiving antenna. Usually moisture buildup in a section of waveguide is the cause. Since the parabolic antenna structure is one of the most commonly used, its gain and performance are fixed by its geometry. Any distortion of the reflector, of course, will cause the gain and received signal to decrease.

The main concern is the final detected signal-to-noise ratio at the base band level. The overall link performance is the amount of noise and distortion added by the communications link. For the video signal, a S/N test at the receiver site should be done using one of the noise-testing sets, such as the TEK 1430 and a waveform monitor, or the new TEK VM-700. This value, for each video signal, should be recorded and compared to the input video S/N at the transmitter site. For telephony there are several overall S/N test sets available on the market. These test sets are also able to measure other parameters such as pilot carrier levels, crosstalk, intermodulation distortion carrier level, and percentage of modulation. The main test for digital signal quality is the bit error rate (BER). Since carrier-to-noise (C/N) ratio for a carrier that is frequency-modulated with a PCM (pulse code modulation) signal will cause the BER to deteriorate, this C/N should be measured. Noise and interference test sets are manufactured by

Hewlet Packard and Tektronix. Since the use of T-1 carriers operating at a high bit rate (1.544 megabits per second) which carries 24 voice-grade channels are commonly used in the United States, special testing equipment is needed to examine these signals properly. The TEK TC-2000 and TC-1000 are multifunction test systems that are used to test the digital signals according to the digital signal protocols used in the telephone industry. Several of the larger instrument manufacturers also manufacture instruments for the European market and make measurements conforming to the European telephone standards.

5.3 Satellite Communications Systems

At first glance, a communications link employing satellite transmission might seem to be a better and more efficient method than multihop (point-to-point-to-point links) microwave radio. This is true, but the method is not without its special problems. Point-to-point microwave was subject to rain fades, antenna icing, as well as corrosion, which gave a link reliability of 99 percent for a conservatively designed microwave path. Satellite transmission also had all of the problems of point-to-point microwave plus one other—an eclipse of the satellite. This occurs twice per year when the satellite for a given spot on the earth's surface is lined up with the sun directly behind it. Since the sun is a large noise generator, the signal-to-noise ratio for the satellite receive signal deteriorates to the point where the signal nearly disappears. This nearly complete outage occurs for only a few minutes and returns slowly to full power at the end of the eclipse period of about 15 minutes at the longest time. This condition limits the path reliability to about 98 percent for the year. Since this condition is predictable, switching the communications traffic to another satellite is the method used by the important and critical communications carriers to get around the eclipse outage time. Communications satellites are placed in a geostationary orbit over the equator at an altitude of approximately 22,300 miles (35,900 km).

The technique used in a satellite uplink downlink is shown in Fig. 5.6. The total signal delay time is nearly one-fourth of a second, which is just about noticeable for two-way (duplex) telephone conversations. One of the more critical problems in construction of a satellite communications facility is the proper pointing of the earth-base antenna structures. Both ground-based antennas have to be pointed exactly at the satellite. Parabolic antennas with a narrow-beam field pattern concentrate the transmit and receive energy to the focal point where the transmission waveguide or the receiving preamplifier is located. Usually a flexible elliptical waveguide connects the transmit energy to antenna structure. At the lower downlink frequency, a high-quality

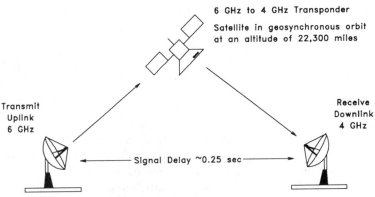

Figure 5.6   Satellite communications link.

coaxial cable connects the received signal to the receivers located in the electronic equipment building. Sometimes this cable was of the pressurized type where dried air was pumped in to keep out moisture, which would cause the signal to decrease. Also, another technique was to convert at the receiving antenna, the downlink frequency band of 3.7 to 4.2 GHz to a lower-frequency band of 950 to 1450 MHz. The coaxial line loss at the receive site was nearly cut in half at this lower signal. By employing the low-noise receiving preamplifier with the downconverter at the receive antenna, the received signal was increased sufficiently to allow a lower gain and less expensive receiving antenna. This trade-off is used in most satellite receiving ground stations.

Naturally the satellite does not stay in its assigned position indefinitely, and periodically the ground controllers have to fire small rocket motors to keep the satellite on-station. Also the on-board rechargeable batteries slowly deteriorate due to the many charge-discharge cycles. These two reasons are the principal ones that limit most satellites to about 8 to 10 years of useful life.

5.3.1 Satellite system parameters and instrumentation

Instruments used to make performance tests on satellite communications systems are nearly the same as those used for the microwave link tests. The carrier levels are tested at the various points in the system using a power meter or a microwave spectrum analyzer. The spectrum analyzer is able to provide carrier level and C/N measurements as well as interference levels from other microwave sources. Point-to-point microwave paths operating in close proximity of a satellite receiving station can cause interference great enough to cause

TABLE 5.1   Satellite Assigned Frequency Bands

Path	Frequency band, MHz
Uplink	5925–6425
Downlink	3700–4200
Uplink	7900–8400
Downlink	7250–7750

some signals to be unusable. The FCC, if notified, will provide information of common microwave carriers operating in the vicinity where a receiving antenna is desired. Several companies provide a frequency-coordinating service to users of satellite receiving systems. This service will tell a communications carrier of existing microwave paths and the expected interference level. Oftentimes a test feed horn or portable parabolic antenna with a preamplifier connected to a microwave spectrum analyzer can be set up at the prospective receive site and an on-site investigation of any measureable interference can be made. If this type of test is made over a reasonable period of time, then any observed interference can possibly be corrected by selecting special antennas, like the horn type, or shielding barriers can be constructed. In the event that interference is too difficult or expensive to correct, another receive site can then be selected.

The usable frequency window for satellite communications is between 1 and 10 GHz. Four 500-MHz-wide bands are assigned for satellite communications, as listed in Table 5.1. These bands are further divided depending on their use. Video program suppliers to cable television systems have 24 channels with each adjacent channel of opposite polarity. Also, adjacent satellites have common frequencies of opposite polarization. Table 5.2 shows the channel plan for two adjacent satellites Satcom 3R (131 W) and Galaxy I (134 W).

Notice that the transponders are 20-MHz apart and at opposite antenna polarization, which reduces adjacent channel interference to below threshold limits for properly designed and installed equipment.

5.3.1.1 Downlink tests and measurements. The polarization of the receiving antenna can be corrected by measuring the desired carrier level with a spectrum analyzer and adjusting the polarization to maximize the desired carrier level while minimizing the adjacent carrier of opposite polarity. This measurement can be made at the 3720 to 4180 MHz band or the block-down converted band of 950 to 1450 MHz. A typical satellite receive-only site for cable television programming delivery is shown in Fig. 5.7. Block conversion is performed within the integrated low-noise preamplifier, one for each pole. The lower frequency is coupled to the power divider through coaxial cable at a

TABLE 5.2   C-Band Satellite Transponder Frequencies

Transponder number	Transmit uplink frequency, MHz	Receive downlink frequency, MHz	Polarization, Satcom 3R	Polarization, Galaxy I
1	5945	3720	V	H
2	5965	3740	H	V
3	5985	3760	V	H
4	6005	3780	H	V
5	6025	3800	V	H
6	6045	3820	H	V
7	6065	3840	V	H
8	6085	3860	H	V
9	6105	3880	V	H
10	6125	3900	H	V
11	6145	3920	V	H
12	6165	3940	H	V
13	6185	3960	V	H
14	6205	3980	H	V
15	6225	4000	V	H
16	6245	4020	H	V
17	6265	4040	V	H
18	6285	4060	H	V
19	6305	4080	V	H
20	6325	4100	H	V
21	6345	4120	V	H
22	6365	4140	H	V
23	6385	4160	V	H
24	6405	4180	H	V

lower loss for each pole. For this satellite, the odd transponder numbers are vertically polarized and the even numbers are horizontally polarized. For the adjacent satellite, the reverse polarization is used. The spectrum analyzer can be connected to the directional coupler test points to check for any cross-polarization problem.

Measurement of carrier-to-noise ratio is often performed on the satellite receiver's 70-MHz IF output. Since this is next to the detector, this value of C/N is useful in determining video quality. A value greater than 51 dB is considered a reasonably good C/N value. This measurement can be made with a power meter or with a spectrum analyzer corrected for bandwidth. Most satellite receivers have an IF output test point at the usual −20 dB value. The power meter or spectrum analyzer can be connected to this test point as shown in Fig. 5.8. If a spectrum analyzer is used, possibly the attenuator will not be needed due to the greater dynamic signal range of the instrument. Also, a correction factor may be needed to correct the analyzer's resolution bandwidth (RBW) to that of the receiver's noise bandwidth. This procedure can be repeated for each satellite channel. However,

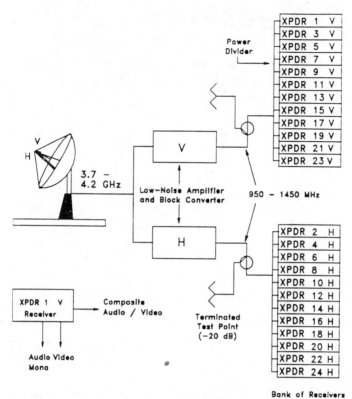

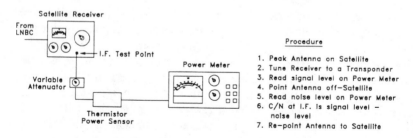

Figure 5.7   Satcom IIIR receive-only diagram.

Figure 5.8   Test for C/N using a power meter.

usually a test at 3720 MHz and 4180 MHz with possibly a channel in the middle is adequate. If a spectrum analyzer is used at either the downlink frequency or the block-down converted band, the relative amplitude of all the transponder channels can be tested as well as any interfering signal carriers. If any terrestrial microwave interference is observed in any of the received channels, the spectrum analyzer can determine the frequency of the interfering carrier and a sharp-notch filter can be placed in the transmission path to attenuate the interfering carrier to tolerable limits. This filter can be placed in the microwave signal path or inside the receiver at IF. The spectrum analyzer screen for one pole is shown in Fig. 5.9*a* and expanded view at Fig. 5.9*b*. Any unidentifiable carriers appearing could be causes of interference. The connection diagram for the spectrum analyzer screens is shown in Fig. 5.10. A measurement of the relative amplitude of the transponder frequencies can be made as well as identification of any interfering carriers. Also, if the opposite-pole transponders are found in between the carriers, readjustment of the antenna polarization will be indicated. For proper polarization for the 24 transponder carriers, the opposite-pole carriers should be minimized for each pole. Peaking the polarization adjustment will maximize the carrier levels, while minimizing the carrier levels belonging to the opposite pole. Peaking the azimuth and elevation angle adjustments will maximize the carrier levels for both poles as read from the spectrum analyzer screen.

5.3.1.2 Uplink tests and measurements. The uplink or transmitting system should be maintained to extremely high standards. Many receiving stations' signal quality depends heavily on the signal quality of the transmitting system. Oftentimes problems with the satellite transponders can be compensated for by making adjustments at the transmitting site. Control of the satellite-positioning system and the on-board powering systems is performed by the satellite manufacturers, who are often the owners. Most transponders are leased by the users, who have control over the on-board transponders and the uplink transmitter. The receiving stations are usually licensed by the lessors of the transponders. Satellites used for telephony and or data transmission operate in a two-way mode and require two on-board transponders, one for the downlink and one for the uplink. A simplified up/downlink diagram is shown in Fig. 5.11. Satellite transmitting systems are typically frequency-modulated due to the noise-improvement performance of FM systems. Typical satellite uplink equipment uses large parabolic antenna structures fed by a high-power amplifier (HPA) operating at 0.5 to 3 kW. This power translates to approximately +57 to +65 dBm. The TEK 492 spectrum analyzer, as well as others, is limited to +30 dBm input power (1 W). Therefore, a −30-dB directional coupler with an added 20-dB attenuator will do the job as

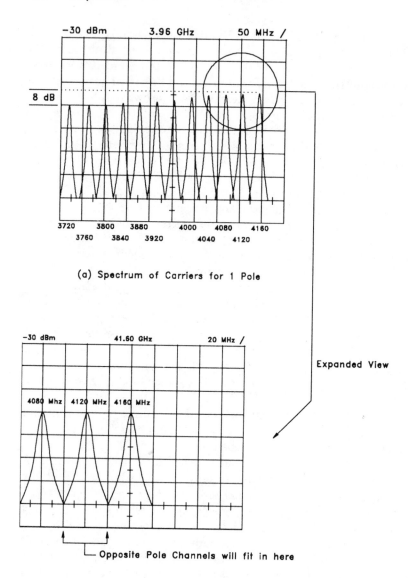

(a) Spectrum of Carriers for 1 Pole

Expanded View

(b) Upper 3 Channels on Same Pole at 20 MHz RBW

Figure 5.9 Spectrum of carriers for 1 pole.

shown in Fig. 5.12. For accurate results, modulation should be turned off and a high-power dummy load in place of the antenna should be used. This dummy load should be equal in impedance match to the antenna system.

To test for proper percentage of modulation, the Bessel null method

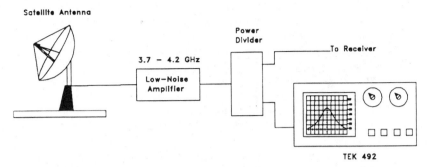

Figure 5.10   Spectrum analyzer test at 3.7 to 4.2 GHz satellite transponder.

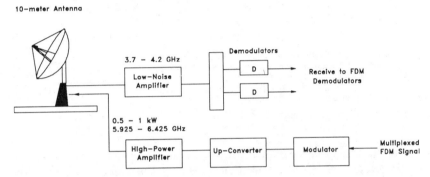

Figure 5.11   Two-way satellite up/downlink station.

can be used and, as is often the case, the 70-MHz IF carrier that will be up-converted to microwave frequencies can be tested. Nulling of this 70-MHz carrier with the proper modulating signal level at the proper modulation frequency will provide the null at the 100 percent modulation point. As explained previously, this test is an out-of-service test. However, use of the TEK 492 "Max Hold" feature, a check of overall carrier deviation ($\pm$ 18 MHz), can be performed. The system connections are shown in Fig. 5.13a and a sketch of the TEK 492 screen showing the test results appears in Fig. 5.13b.

5.3.2   End-to-end testing

Certainly the "acid test" of any communication system is to compare the received signal quality to the transmitted signal quality. The accumulated noise and distortion components always cause the received signal to be of lower quality than the transmitted signal. The more pure the received signal, the better the communications link performs. All of the foregoing tests have a direct bearing on the link

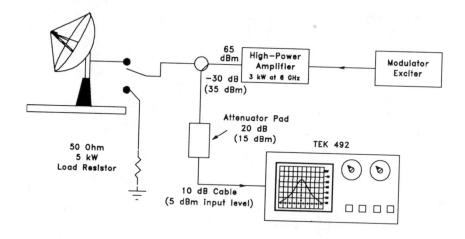

Loss from Amplifier to Spectrum Analyzer is 60 dB. If Spectrum Analyzer indicates a signal level of 5 dBm, the High-Power Amplifier will be delivering 3 kW (65 dBm) to the Antenna feed with a loss of 3 dB. If the Antenna has a gain of 50 dB, then the Effective Radiated Power (ERP) is 112 dBm or 158 Megawatts.

Figure 5.12 Uplink power measurement.

performance, and in case of difficulty will often point to the defective circuit or component. Continuous monitoring of the signal quality with periodic measurements recorded in a logbook is extremely helpful and will often provide a warning that a circuit or component element is failing before a catastrophic failure occurs. For any communication link, signal quality and reliability are the most important criteria.

5.3.2.1 Television round-trip quality tests.
For one-way links such as those satellite facilities distributing television programming to cable television systems, a video test signal inserted in each video signal during the vertical interval can be used to measure the parameters governing video signal quality. This procedure was developed to ensure that the microwave radio links connecting the television broadcasting networks from point to point across the country were performing to the necessary standards. These test procedures and standards were published in a report called NTC-7 (Network Transmission Committee No. 7). It should be recalled that such testing procedures were covered in Chap. 4, Sec. 4.2.2.

When testing a two-way analog link such as for television network distribution, a so-called round-trip test can also provide a continuous signal-monitoring program. The VITS signal can be generated at one end and transmitted to a receiving station where this signal is re-

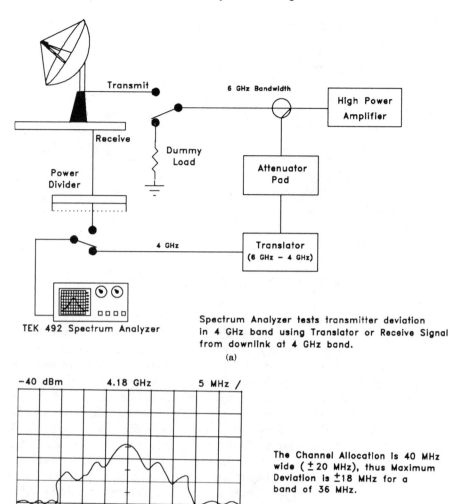

TEK 492 Spectrum Analyzer

Spectrum Analyzer tests transmitter deviation
in 4 GHz band using Translator or Receive Signal
from downlink at 4 GHz band.

(a)

The Channel Allocation is 40 MHz
wide (± 20 MHz), thus Maximum
Deviation is ±18 MHz for a
band of 36 MHz.

Center
Freq.

Figure 5.13   (a) Transmitter deviation test; (b) spectrum analyzer screen of carrier
deviation test.

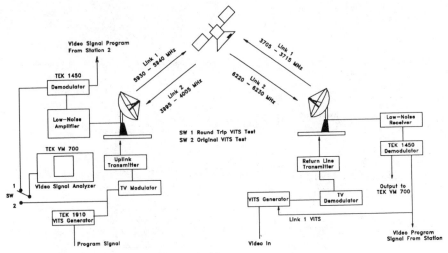

Figure 5.14   Round-trip vertical interval video test.

turned on a downlink path back to the sending station. This signal quality can be continuously monitored for the types of signal distortion and noise contributions of the round-trip signal path. The main problem is to determine which path is at fault in the case of measurable signal degradation. The uplink can be checked out by having someone make measurements at the receiving end and phone the results to the transmitting end. If this path checks out, the fault may be in the return path and someone at the receive site can break the link and transmit back a pure signal to the first transmitting site. This type of test procedure is shown in Fig. 5.14. Essentially the original VITS signal is transmitted by station 1 through link 1 to station 2 and is reinserted into the video signal at station 2, to be returned to station 1 via link 2. This round-trip VITS signal tested at station 1 performs an overall loop test for signal quality.

5.3.2.2 Digital modulation tests.   Digital signal testing involves use of specialized equipment designed exclusively for digital signals. The TEK 1200 series logic analyzers can be used to test the demodulated digital signal. Since the bit error rate (BER) is an extremely important parameter, many such instruments are on the market today from several manufacturers.

For people with sophisticated communications systems and good-size equipment budgets, a complete signal-monitoring system might be justified. Combinations of test equipment can be placed together to make up a specialized test system. Such a system is shown in Fig. 5.15. The 18-GHz signal is split off through a directional coupler to the

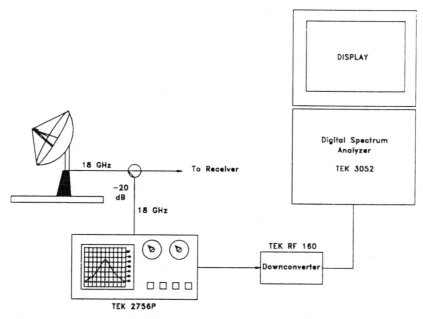

Figure 5.15   Satellite downlink RF testing of digital data.

TEK 2756P microwave spectrum analyzer, which when tuned to this 18-GHz signal can be used to analyze the side-band structure of the carrier. The 110-MHz IF output signal from the spectrum analyzer can be down-converted to base band by the TEK RF160. This base-band signal can be thoroughly analyzed by the TEK 3052 digital spectrum analyzer. Also, this system can be programmed by a computer and the data analyzed by appropriate UNIX-based software for complete automatic testing and signal monitoring. If FSK signals are used, the TEK S2MG100 enhanced-frequency-measurements software is more appropriately used with the TEK 3052.

5.4   Radar Systems

Radar systems were one of the earlier users of microwaves due mainly to the more precise line-of-site characteristics of microwave propagation. The name *radar* was derived from "radio direction and range." A radar system consists of a pulsed RF transmitter coupled to the usual parabolic antenna structure. This pulse of RF microwave energy travels in a narrow beam striking a metallic target in the beam's path. Energy is reflected from this target and is received by the parabolic narrow-beam antenna. The time it takes for the pulse to make the round trip is a function of the distance between the transmitter/

receiver antenna and the target. The transmitting and receiving equipment is colocated, that is at the same site, and share a common antenna system.

There are two types of radar systems. One system follows the basic form of measuring the distance and direction of an object or target. The second is called the *doppler type,* which measures the relative motion between the radar site and the target. One should recall that the doppler effect is the frequency shifting of the transmit/receive carriers due to the relative speed between the radar set and the target. Motor vehicle speed measurement by law-enforcement officials is a form of portable doppler-effect radar. Also, the military uses this type of radar in aircraft and on board ships as well as in the field. Military-type radars are usually maintained, tested, and calibrated by specialized test equipment made for that purpose. In most cases, test ports and/or test umbilicals are provided to connect on to the test equipment, which in many cases is programmed by a computer or microprocessor. Long-range radars of the type used for early-warning air defense also are highly specialized as well as their maintenance, testing, and calibration procedures. This section will be mainly concerned with the basic type of radar similar to that used on board merchant ships to avoid collisions and to aid in navigating in fog and bad weather.

5.4.1 Marine Radar Systems

Commercial marine radars operate through frequency bands allocated specifically for radar use. One commonly used marine band is the 3000 to 3246 MHz band, often referred to as S-band. A full-wave antenna at this frequency band is about 10 cm. Another commonly used band is X-band at 9320 to 9500 MHz with an approximate wavelength of 3 cm. The third band at 5460 to 5650 MHz is also available for radar use. Rotating-slot antennas or parabolic reflector antennas are used to produce a typical beam width at 2 to 3 degrees (horizontally) and 15 to 20 degrees vertically. Most marine radar antennas are mast-mounted and are rotated continuously at about 10 r/min. In many instances, the transmitter/receiver set is placed in an enclosure mounted at the rear or just below the scanning antenna structure. This makes the waveguide length short, thus reducing losses and making for a more sensitive and efficient radar. Unfortunately, it means one of the technicians may have to climb the radar mast for routine lubrication and maintenance checks.

Most marine or ship-board radars provide target information from about 100 yards from the ship to nearly 100 nautical miles. A nautical mile is 6080 feet (USA) where a statute (land-based) mile is 5280 feet. Radar waves travel at 300,000 m/s, which is approximately 186,000 statute miles per second or 162,000 nautical miles per second. Therefore, in 1 μs a wave will travel 0.162 nautical mile or 1 nautical mile in 6.17 μs. The round-trip time from the transmitted pulse to the received echo pulse the wave has to travel 12.3 μs for a nautical mile. Therefore, 123 μs corresponds to 10 nautical miles.

5.4.1.1 Marine radar set operation. A block diagram for a simple marine-type radar system is shown in Fig. 5.16. The function of each block will be discussed. The timer modulator generates the basic transmitted pulse repetition rate (PRR), which may vary from 800 to 2000 pulses per second as selected by the range switch on the indicator console. A lower PRR is selected for longer ranges as well as a longer RF pulse duration. This increases the average power output of the ra-

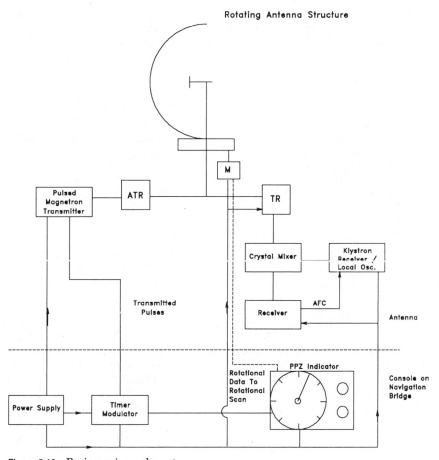

Figure 5.16   Basic marine radar set.

dar transmitter. Since the radar technique is to send out a high-power short pulse of RF power and then to receive for a longer transmitter off time for the echo. A form of switching system is required to prevent the high transmitted pulse from burning out the sensitive receiver circuits. The TR (transmit/receive) box performs this function. The TR box is a tuned resonant cavity containing a special glass gas-filled tube and a spark-gap set of electrodes. A high dc voltage is placed across the spark-gap electrodes. When the magnetron fires, a high-power pulse causes a high-voltage pulse to appear across the spark-gap electrodes, which, with the aid of the bias voltage called the keep-alive voltage, arcs across causing a short-circuit condition and preventing any energy to feed into the receiver. All the transmitted energy continues on to the antenna. Once the pulse is over, no arc appears across the receiver, thus allowing the receive echo energy to enter. The ATR box (anti-transmit/receive) is placed one-quarter wavelength from the TR box near the magnetron. When the magnetron fires, this tube also fires, allowing the transmitted energy to pass through to the antenna. When the pulse of energy ends, the tube deionizes, causing a short circuit across the waveguide, which now reflects any received echo toward the receiver, thus preventing any of the received signal into the magnetron structure. This whole waveguide feed assembly is often referred to as the *duplexer*. The local oscillator (klystron) for the receiver feeds energy at usually a 30-MHz difference from the transmit/receive frequency to the crystal mixer, also built into the duplexer. The output of this mixer is fed to a multistage high-gain 30 MHz (IF) amplifier. The bandwidth of this amplifier has to be 10-MHz wide in order to pass the sidebands caused by the sharp narrow transmitted pulse. An AFC (automatic frequency control) voltage is developed in the receiver to control the klystron frequency to maintain the 30-MHz IF difference frequency. The reflex klystron can have its frequency changed over a narrow band by controlling the voltage on the repeller electrode. Also, the magnetron frequency can be changed a small amount by adjusting the magnitude of the negative-voltage pulse on the cathode.

The indicating system consists of a cathode-ray tube, with a diameter that may vary between 7 to 15 inches, mounted in a console on the navigating bridge. A bright trace forming a radius line on the CRT is rotated at the same speed as the rotating antenna structure. When this trace is at the 12 o'clock position (straight up), this position corresponds to the antenna structure aiming at the ship's bow (heading). Every time this trace is swept through this position, the brightness is increased. This is called the *heading flasher*. The general brightness level is set so the rotating trace cannot be quite visible. The echo signal acts to brighten this rotating trace causing a map or picture of the surrounding objects to be displayed on the CRT. A scale in degrees surrounds the CRT display, which can be rotated to line up with the echoes. This can provide directional (azimuth) information from the target echo and the ship's heading. Since the length of the radius line represents the maximum distance on the range-scale setting of the radar, a short brightening pulse will provide a ring on the display, which can correspond to a distance measurement. A servomechanism provides the rotation of the CRT deflection coil to correspond to the antenna rotation. The indicating CRT has a fairly long persistence (approximately 10 s), which can hold the image until the trace comes around another circle. Oftentimes range rings can be selected for 1, 5, or 10 nautical miles on the 10-mile range and 10, 25, and 50 on the 50 nautical mile scale. This type of display is often referred to as a *PPI (plan position indicator) scope*. The receiver has a sensitivity control for aid in identifying strong and/or weak targets. A sea return control, when used, can eliminate or reduce close-in echos caused by a turbulent sea near to the ship without reducing the longer-range sensitivity. Of course, a close-in large echo, caused by a nearby ship, will override the sea return control.

5.4.1.2 Marine radar parameters. Radar range depends on the power output of the transmitter and the sensitivity of the receiver. The receive echo signal, in watts, is given by the following mathematical expression:

$$P_r = P_t Ae^2 \sigma / 4\pi\lambda^2 R^2$$

where P_r = the received echo power, in watts
P_t = the transmitted signal power, in watts
Ae = antenna effective area, in square meters
σ = the radar cross section of target, in square meters
R = range from radar set to target, in meters
λ = wavelength, in meters

The value P_r is the available signal power presented to the receiver for detection. If this signal is too small, the receiver will never be able to distinguish it from the accumulated noise. It should also be clear that for the range to double, the transmitted power has to increase 16 times. Also, the PRR has to be long enough for distance echos to be detected. As previously stated in Sec. 5.4.1.1, the PRR usually is in the range of 800 to 2000 pulses per second. These pulses have a duration typically of 0.25 to 1 μs at a high peak power of 15 to 25 kW for commercial marine radar. Since the duty cycle is long, the average power

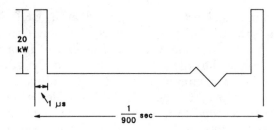

Each second contains 900 x 1 x 10^{-6} = .0009 second pulse. The transmitter is "on" .0009 seconds for each second. Therefore, the duty cycle is .0009 for the peak power of 20 kW, giving an average power of 20,000 x .0009 or 18 Watts. Since magnetrons are about 50% efficient, the input power must be at least 36 Watts.

P AV = P PEAK x PULSE WIDTH x PRR   since Pulse Width x PRR = Duty Cycle
P AV = P PEAK x DUTY CYCLE

Figure 5.17 Pulsed radar relationship.

is actually quite low. An example is shown in Fig. 5.17. From this example, since the average magnetron RF output is 18 W, a magnetron at 50 percent efficiency will draw a measured magnetron power of 36 W.

5.4.2 Radar system maintenance

Radar systems require a rigorous maintenance program to assure system accuracy and reliability. For marine use, radar is necessary for safe navigation and marine safety. The Federal Communications Commission requires radar maintenance personnel to be licensed with a specific marine radar endorsement attached. An examination is one of the requirements for both the license and endorsement. Maintenance and repair personnel for aircraft navigational radars also require a special endorsement on the general class radio and telephone license.

5.4.2.1 Radar testing precautions.

Before performing maintenance or repairs on radar systems, some precautions should be observed. Since radar equipment requires high voltages necessary to produce the high-power RF pulses, most equipment enclosures have interlocks that, when the cabinets are opened for maintenance, the power is disconnected. Such voltages fall in the range of 10 to 20 kV, so extreme caution should be observed. Most high-voltage power supplies contain filter capacitors that will maintain a high voltage charge. Some radars have a discharge switch that discharges the capacitors when the transmitter cabinet is opened. A careful reading of the radar instruction manual should be done before performing any maintenance procedures, and such capacitors should be short-circuited with insulated clip leads.

The high-RF pulse power of a radar transmitter can cause serious burns to anyone operating close to the antenna structure. The higher power the radar, the more serious the burn. Also, since the antenna structure rotates, caution should be taken to avoid getting hit or injured by the antenna.

The CRT in the indicator section of the radar can cause a hazard, since the glass envelope has a high vacuum. If this tube is hit hard enough, it will implode, causing flying glass which can cause a serious injury.

The magnet structure surrounding the magnetron is very powerful and can damage instruments as well as a number of watches. Caution should be observed when working near this magnet. It is a good idea when working on radar equipment to remove jewelry.

5.4.2.2 Radar tests and measurements.

Most radar systems have various test points placed in the circuits where instruments can be connected to monitor important operating parameters. Such instruments may be a digital multimeter or an oscilloscope. Crystal current, magnetron current, klystron current, AGC (automatic gain control) voltage, and power-supply voltages and currents are some of the parameters. The pulse waveforms are usually analyzed using a high-quality oscilloscope such as the TEK 2000 series. Good oscilloscope synchronization and the ability to see and analyze short narrow pulses are required. Precise pulse timing is necessary to make the accurate range measurements required of most radars. Use of the delay sweep capability available on many oscilloscopes aids in examining the leading edge of the modulating pulse. A directional coupler placed in the waveguide, or from a small antenna placed near the reflector, will allow an examination of the pulse at microwave frequencies using a spectrum analyzer. The connections and a typical screen graticule for a proper pulse spectrum are shown in Fig. 5.18. Caution should be observed when connecting a spectrum analyzer to a directional coupler. For the TEK 492 the maximum allowed signal is +30 dBm and for higher-power radars, an attenuator will be needed at the spectrum analyzer input.

A handy testing device which gives an overall check of a radar system is called an *echo box*. This device is simply a tuned cavity coupled

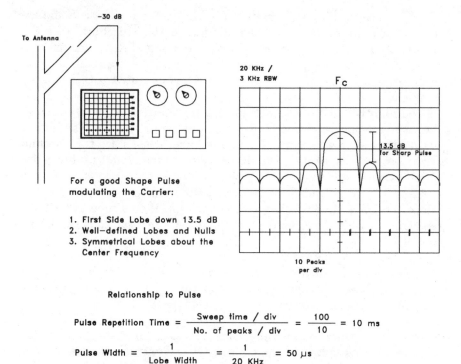

For a good Shape Pulse
modulating the Carrier:

1. First Side Lobe down 13.5 dB
2. Well-defined Lobes and Nulls
3. Symmetrical Lobes about the
   Center Frequency

Relationship to Pulse

$$\text{Pulse Repetition Time} = \frac{\text{Sweep time / div}}{\text{No. of peaks / div}} = \frac{100}{10} = 10 \text{ ms}$$

$$\text{Pulse Width} = \frac{1}{\text{Lobe Width}} = \frac{1}{20 \text{ KHz}} = 50 \text{ } \mu\text{s}$$

Figure 5.18 Pulse sin x/y envelope on spectrum analyzer.

to the waveguide feeding the antenna. Every time the transmitter fires a pulse, the cavity will resonate at the transmitter frequency and the oscillations will decrease with time, causing the receiver to detect these signals as echos that get weaker. These decreasing echos last from 10 to 12 μs. A weak receiver or failing crystal mixer will only see the close-in or main echos. Recall that 12 μs corresponds to 1 nautical mile of range. Therefore, a good receiver and crystal mixer should see these echos out to about 1 mile. Most echo boxes have a tuning control, so if the box is left in the directional coupler, it can be detuned from the carrier frequency when not in use. This is a fairly low cost and simple device that can provide an overall test of the system.

If the crystal mixer is suspected to be a problem, it can be removed and tested with a simple ohmmeter front-to-back ratio test. With the ohmmeter on the R × 1000 scale, a 10 to 1 or greater front-to-back ratio will indicate that the crystal is good. Caution should be observed when handling this crystal device because accumulated static electricity can ruin the sensitive microwave crystal. A front-to-back crystal check is shown in Fig. 5.19. Typically, the backward resistance will be

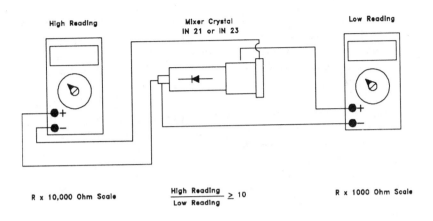

High Reading   Mixer Crystal IN 21 or IN 23   Low Reading

R x 10,000 Ohm Scale $\dfrac{\text{High Reading}}{\text{Low Reading}} \geq 10$ R x 1000 Ohm Scale

IN 21, IN 23 Crystal Nearly Full Size as Shown

Figure 5.19   Crystal front-back ratio resistance test.

about 10 to 100 kohms, and the forward resistance between 100 and 1000 ohms.

Since the magnetron and klystron are vacuum tubes, they require filament power that is usually supplied by a separate filament ac transformer. The magnetron filament is usually connected to the cathode and the filament transformer has to be heavily insulated to withstand the high negative-pulse voltage used to fire the magnetron oscillator. The magnetron filament supply has to be activated before the magnetron can be pulsed or the magnetron can be damaged. Usually a delay mechanism prevents the magnetron from being pulsed until the filament has been allowed to reach proper operating temperature.

5.4.2.3 Radar system diagnostics. Some of the more frequent problems many commercial radar systems experience can be isolated to either the transmitter, receiver, indicator-timer, or the main power supply. Some smaller marine radars are designed to operate from dc power derived from a marine battery supply. This type of radar may utilize what is known as a dc to ac converter, where the battery supply is converted to alternating current and distributed to the various circuits. In many cases, rectifiers in the various sections such as the receiver, convert the alternating current back to a desired dc level.

Failure of the mixer crystal, or as is often the case a pair of crystals, is quite common. Since a failing TR or anti-TR box can cause crystal deterioration, replacement of the TR and of the anti-TR tubes as well as the crystals is the usual procedure. Crystal failure can often be indicated when only larger close-in targets are observed. Oftentimes the klystron receiver local oscillator may need some tuning adjustment, and usually the AFC is switched off and the tuning adjustment peaked for maximum distance echos. Then the AFC circuit is switched back on and the distance echos should remain.

Since most maintenance procedures and recommendations are thoroughly discussed in the system manuals, a periodic reading is highly recommended. The complete radar system has a lot of electromechanical components that require cleaning and lubrication. Such procedures should be described in the manuals and rigorously observed.

6

Cable Communication Systems

6.1 Cable Systems

Cable systems, as most of us know, consist of a network of metallic wires used to distribute electrical power, telephone service, cable television services, and lately computer-related distributed systems. This network of wires used for communications and/or communication systems is the concern of this section. Lately, the use of fiber-optic cables has been integrated into this communications network. It may seem strange to some that the main concern formerly was to provide a low-resistance path through the network to allow the electrical energy to flow and now, here we are, sending electrical energy through a fiber of glass, an insulator. As will be made evident, it is the frequency of the alternating power that allows this phenomenon to take place. The metallic cables consist of twisted wire pairs, with one or more twisted pairs bundled in a cable with a metallic foil shield and encased in a polyethylene jacket; or the cable can be of the coaxial variety where a metallic-center conductor is suspended in a thin-wall metallic tube filled with insulating plastic foam. The thin-wall tubing, made usually from aluminum alloy, may be covered with a protective polyethylene jacket. This type of cable is used for cable television systems or for commercial or industrial local area communication networks.

6.1.1 The twisted wire pair

The twisted pair type of cable is the mainstay of the telephone industry. Since the telephone industry is concerned with voice frequencies in the band of 300 to 3400 Hz as well as low-speed telegraph-type signals, the twisted-pair cable was adequate. Also, it was able to stand up under the stresses of utility pole installation as well as being placed in underground buried conduits. Older-type telephone cable consisted of copper wire pairs insulated with a fabric covering of silk or cotton. These twisted pairs were in turn twisted together, covered with wax paper, and encased in a seamless lead jacket. When this lead jacket split due to stress and temperature or even animal gnawing, moisture would permeate the cable, causing telephone service to become cross-connected. Since a telephone conversation could be heard as a background in another pair, such leakage from one circuit to the other was known as crosstalk. The cure was to cut out the section of damaged lead, spread the pairs out and pour hot wax to further insulate and dry the bad section. A sleeve of lead was placed over the section and the section was carefully sealed with molten solder. Highly skilled personnel were required to work on lead cable repairs and splicing. Modern telephone cable still uses copper wires, but high-quality plastic insulation, cable outer jackets, and self-venting splice enclosures have increased plant reliability problems and the ability to withstand the weather and pole plant environment.

6.1.1.1 Telephone wire systems. The telephone cable plant has matured to the point that POTS (plain old telephone service) is one of the most reliable communication systems in existence today. This has come about due to the dedication the telephone industry has for reliability. In the Second World War, the telephone industry specification was adopted almost completely by the Army Signal Corps. Hence, at the war's end many well-trained personnel released from the Signal Corps were available to the telephone industry.

Still, the usable frequency bandwidth of twisted-pair cable as compared to other types of cable is small. Voice frequencies that transmit the necessary intelligibility range from 300 to 3400 Hz. However, the twisted pair can be used at frequencies of nearly 110 kHz. However, the cable attenuation increases with increasing frequency. Therefore, for frequencies up to 100 kHz, twisted-pair losses are huge. Open wire line can carry higher frequencies. Twisted-pair cables have to have points of equalization where passive networks are placed to flatten the frequency response. For certain applications the telephone company will provide equalized conditioned lines to some customers. Such customers may need this service for computer network uses or radio stations for program feeds from studio to transmitter. Since the telephone industry has established proper maintenance and test procedures with customized test equipment that would fill several volumes only basic test methods for a so-called in-house network will be discussed. Generally, for one twisted pair of copper wires the shunting capacitance per kilometer is in the order of 40 to 50 nF (nanofarads). However, for 26-gauge copper wire the loop resistance is 270 ohms/km with an attenuation of 1.6 dB/km. Cable loop resistance can be tested by simply short-circuiting the pair on one end of a known length of cable and measuring across the other end with an ohmmeter. If more accuracy is required, a Wheatstone or Kelvin bridge can be used. To measure attenuation, a signal source of ac voltage can be placed at one end and a meter at the other end. By measuring the input power from the generator and the power at the receiving end, the attenuation can be measured. At 1000 Hz, the wavelength calculates to 300 km, so one-fourth of a wavelength is 75 km. Measurements at 1000 Hz should be treated as nearly dc. Measurements of loop resistance and attenuation are shown in Fig. 6.1a,b. These are the tests one would make to confirm the parameters. However, if the cable size and length are known, the parameters may be calculated from the resistance per unit of length given in wire tables. An example is shown in Fig. 6.2. The typical 600-ohm source load impedance of most telephone equipment is used in the example. For normal small-sized systems under 5 km (15,000 ft), the accumulated capacitance does not pose a problem with frequency response in the voice range. Longer systems require series inductances placed at intervals along the line. The telephone industry has worked out the loading procedure of long lines using standardized values of loading inductances spaced at specific intervals. The typical U.S. telephone subscriber circuit consists of a pair of wires plus a ground return path with a 48-V dc power source, which supplies power to the carbon microphone telephone transmitter. The handset receiver

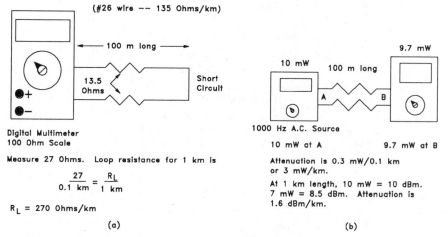

Figure 6.1   Loop resistance and attenuation for a twisted-pair cable type.

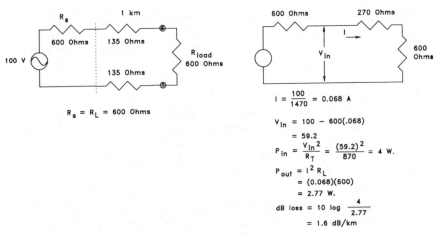

Figure 6.2   Circuit calculations for a twisted-pair loop.

is an electromagnetic earphone. The ringing power is a 20-Hz 90-V ac source which causes the telephone set bell to ring. The subscriber loop resistance may vary from 0 to 1300 ohms with a maximum loop loss of 8 dB. Usual telephone systems conduct only a small amount of current on the order of 20 to 80 mA. To provide proper voice intelligibility, the receive sound (earpiece) pressure level ranges between 70 and 90 dB spl (sound pressure level) with a distortion limit of −50 dB. The ring power of 90 V at 20 Hz ac and the −48 V dc are supplied by the telephone utility plant.

One technique used and founded by the telephone company is the dual-tone multifrequency (DTMF) dialing method used on touch-tone telephone sets. The old dialing technique was derived from a serial pulse generator called a rotary dialer. By placing the finger in one of the dialing holes and rotating the ring to the stop caused the loop circuit to be interrupted by the number corresponding to the dial hole. To speed up the dialing process, the touch-tone DTMF system was developed. DTMF signals are used to control various electromechanical devices such as tape recorders, switching equipment, and security systems. These tones are transmitted through telephone systems and radio and television broadcasting, microwave, and satellite communications systems. Several manufacturers make integrated circuit chip sets for DTMF generation and decoding. The DTMF system is shown in Table 6.1.

It should be evident from the table that 12 selections can be made with 7 different frequencies operating in a pair as selected by pressing a single button. For instance, pressing the number 5 (JKL) button, the two tones generated are at 770 and 1336 Hz. All of these tones appear in the normal voice band of frequencies of 300 to 3400 Hz. Some tele-

TABLE 6.1   DTMF Keypad Frequency Pairs

Number/letters	Tone 1, Hz	Tone 2, Hz
1	1209	697
2 ABC	1336	697
3 DEF	1477	697
4 GHI	1209	770
5 JKL	1336	770
6 MNO	1477	770
7 PRS	1209	852
8 TUV	1336	852
9 WXY	1477	852
*	1209	941
O OPER	1336	941
#	1477	941

phone answering machines are controlled by the calling touch-tone telephone. For example, a salesperson can call his or her home and reach the answering machine and then by further use of the telephone keypad instruct the answering machine to play back the messages. Therefore, the answering machine must be able to decode and react to the DTMF tones.

6.1.1.2 Telegraph wire systems. The twisted-pair lines can also be used to carry low-frequency and/or low-speed digital signals. The earliest use of digital electrical signals was the Morse code telegraph. Essentially a voltage was pulsed on a pair of wires by a battery and telegraph keying device. The dots and dashes were keyed by an operator. The success of this system depended on the expertise of the operator using the key and listening to an electrical solenoid sounder, or clicker, as it was often called. Each station had its number, and several stations would be connected to a line. Since the early telegraph systems were built and owned by the railroad industry, commercial use for domestic messages was an outgrowth. Most of the pole-line telegraph plant was built on the cleared railroad rights-of-way. The early telegraph systems used open-wire line on glass insulators mounted on utility pole cross arms. A busy railroad right-of-way usually had several cross arms with many glass insulators supporting the copper telegraph wires. These lines were mostly resisitive since the accumulated capacitance and inductance were small. Once this service was condensed into multiple twisted pairs, problems in telegraph service caused the industry to use a system of tones or modulated tones.

As it exists today, telegraph consists of what is known as teletype. The old method of sending Morse code has been replaced by

7 6 5 4 3 2 1	Row	0	1	2	3	4	5	6	7
		0 0 0	0 0 1	0 1 0	0 1 1	1 0 0	1 0 1	1 1 0	1 1 1
0 0 0 0	0	NUL	DLE	SP	0	@	P		p
0 0 0 1	1	SOH	DC1	!	1	A	Q	a	q
0 0 1 0	2	STX	DC2	"	2	B	R	b	r
0 0 1 1	3	ETX	DC3	#	3	C	S	c	s
0 1 0 0	4	EOT	DC4	$	4	D	T	d	t
0 1 0 1	5	ENQ	NAK	%	5	E	U	e	u
0 1 1 0	6	ACK	SYN	&	6	F	V	f	v
0 1 1 1	7	BEL	ETB	'	7	G	W	g	w
1 0 0 0	8	BS	CAN	(	8	H	X	h	x
1 0 0 1	9	HT	EM	)	9	I	Y	i	y
1 0 1 0	10	LF	SS	*	:	J	Z	j	z
1 0 1 1	11	VT	ESC	+	;	K	[	k	{
1 1 0 0	12	FF	FS	,	<	L	\	l	/
1 1 0 1	13	CR	GS	–	=	M	]	m	}
1 1 1 0	14	SO	RS	.	>	N	^	n	~
1 1 1 1	15	SI	US	/	?	O	_	o	del

Figure 6.3 ASCII code.

teletypewriting (TTY) equipment using a commonly accepted code system called the American Standard Code for Information Interchange (ASCII). Most computer operations also use this code for driving printers, typesetting machines, as well as for computer keyboard control. In the basic ASCII code system, a binary zero and one identify the letters with the first 4 bits. Other typing control functions require an additional 3 bits to complete the full ASCII code word of 7 bits. This code is shown in Fig. 6.3. Columns 2 to 7 are quite commonly known, except SP which stands for (space) and DEL (delete). The abbreviations in Columns 0 and 1 are given in Table 6.2.

The method used to transmit the binary zeros and ones making up this 7-bit code word may vary according to the line (wire) quality and distance. In order for good transmission quality, the bits have to be well defined and easy to pick out of the usual noise. The signal type that is often used to transmit ASCII digital data is shown in Table 6.3.

The speed of the bit stream through a set of wires determines the rate of transmission of the characters. Since many TTY systems are electromechanical, they have a maximum operating speed. So for active on-line data systems, the rate of transmission should be below the maximum operating speed of the receiving device. Oftentimes punched-paper tape containing ASCII code words is used to transmit

TABLE 6.2 Definitions of ASCII Abbreviations

Column 0 abbreviation	Definition	Column 1 abbreviation	Definition
NUL	All characters zero	DLE	Data link escape
SOH	Start of heading	DC1	Device control
STX	Start of text	DC2	Device control
ETX	End of text	DC3	Device control
EOT	End of transmission	DC4	Device control
ENQ	Enquiry	NAK	Negative Acknowledge
ACK	Acknowledge	SYN	Synchronous idle
BEL	Attention alarm	ETB	End/transmission block
BS	Backspace	CAN	Cancel
HT	Horizontal tab.	EM	End of medium
LF	Line feed	SS	Start/special sequence
VT	Vertical tabulation	ESC	Escape
FF	Form feed	FS	File separator
CR	Carriage return	GS	Group separator
SO	Shift out	RS	Record separator
SI	Shift in	US	Unit separator

TABLE 6.3 Signal Types for Binary Code Transmission

Signal type	0 digit	1 digit
DC current	No current	Positive current
Polarity/current	Negative current	Positive current
AM	Tone off	Tone on
FM	High frequency	Low frequency
Phase modulation/phase reference	Opposite phase to/ reference phase	Reference phase
Differential phase	Phase inversion	No phase inversion
Perforation (punch cards, tape)	No perforation	Perforation

messages at a rapid rate through a cable to a receiver that stores the words for later printing. Such data storage can be magnetic tape or solid-state memories. The TTY bit speed may vary between 45 bits per second to 150 bits per second, which corresponds to 60 and 150 words per minute, respectively. One common type of FM carrier modulation is called *frequency-shift keying* or FSK. This type of modulation is where the carrier is shifted in frequency between two points corresponding to the one-zero condition. If instead of shifting the frequency the phase of a carrier is shifted 180° corresponding to the one-zero condition, phase modulation results. This is also a common method used in transmitting TTY signals.

6.1.2 Telephone line equalization and loading

Twisted-pair lines can be used at higher frequencies for shorter distances if series inductances are added to compensate for the shunting capacitances between the pairs. Lines can be tested for noise crosstalk and frequency response before being selected for special use. The telephone companies can usually find a wire pair connecting the points of service by searching for the shortest path and the least switch points. The shortest path usually will need the least or possibly no equalization or loading. The telephone industry usually has amplifiers that are able to increase the signal level and correct the response for the next cable segment.

6.1.2.1 Series inductance loading. Series inductances added at intervals along the wire pair will improve the high frequency response. Only so much frequency compensation can be done with these series inductances. Often the spacing can be decreased as well as the inductance value of each coil, thus approaching the case of a smooth line. When testing a compensated line for response, many values of single frequency measurements across the band should be performed or use of a sweep system should be made. The line should be tested first, then any improvement can be easily seen by the addition of any corrective inductances. Lines conditioned in this manner improve the audio frequency response for voice communications or audio program feeds from broadcast studio facilities to the transmitter.

6.1.2.2 Conditioned lines. Conditioned lines can improve the baseband TTY pulse-carrying capabilities as well as the higher frequency-modulated carriers. Since the sharp leading-trailing edges of pulses help define the pulses, good high-frequency response may be required by the receiving equipment. For twisted-pair cable, only so much can be done to make distortion-free smooth transmission lines.

6.1.3 Coaxial cables

Coaxial cables have very wide frequency response compared to twisted wires. Coaxial cables are manufactured in a variety of impedances, with the most common at 50 and 75 ohms. As the name implies, coaxial cable has a center conductor, usually of copper, and an outer concentric conductor of copper or aluminum. A dielectric compound of solid polyethylene or polyethylene foam fills the space, thus keeping the center conductor in the center and outer conductor surrounding it in the concentric condition. A plastic vinyl or polyethylene jacket provides environmental protection. The geometry and the dielectric con-

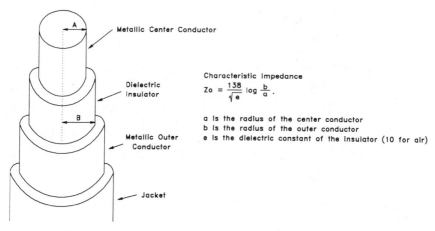

Characteristic Impedance

$$Z_0 = \frac{138}{\sqrt{e}} \log \frac{b}{a}.$$

a is the radius of the center conductor
b is the radius of the outer conductor
e is the dielectric constant of the insulator (10 for air)

Figure 6.4   Coaxial cable constants.

stant of the insulating material supporting the conductors determines the characteristic impedance of the cable. Fig. 6.4 shows the relationship. The dielectric constant for dry air is the reference and is given the value of unity (1). This constant for some other insulating materials is given in Table 6.4.

By foaming the polyethylene or polystyrene, the dielectric constant can approach that of air, since dry air is trapped in the bubbles of foam.

From transmission-line theory, one has to match both the sending end and receiving end impedances to the line's characteristic impedance. The matching of impedances minimizes reflections on the line and allows for maximum power transfer from the transmitter to the receiver. A coaxial cable transmission line has resistance, mostly a series resistance of the conductors, a shunting capacitance formed between the center and outer conductor, a leakage resistance through the insulating material, and a series inductance from the conductors. These constants are usually given on a per hundred feet of cable or on

TABLE 6.4   Dielectric Constants of Materials

Material	Dielectric constant
Free space	1.00
Teflon	2.10
Polyethylene	2.25
Polystyrene	2.55
Nylon	3.40
Quartz	3.78
Glass	4.00

TABLE 6.5   Common Coaxial Cable Specifications

Cable	Characteristic impedance	Capacitance, pF/ft	Attenuation (dB/100 ft), MHz						
			10	50	100	200	400	1000	4000
RG6/U	75	20	0.61	1.50	2.10	3.10	4.40	7.10	14.2
RG8/U	52	21	0.67	1.50	2.20	3.20	4.60	9.00	18.0
RG11/U	75	21	0.41	1.00	1.50	2.20	3.30	5.60	11.2
RG58/U	50	29	1.20	3.10	4.80	7.00	10.00	18.0	NA
RG59/U	75	21	1.00	2.10	2.90	4.10	6.60	11.0	NA
RG214/U	50	30	0.66	1.50	2.20	3.20	4.60	9.00	18.0

a kilometer basis. The constants for some often used coaxial cables are given in Table 6.5.

As can be seen, the loss varies approximately between the frequencies as the square root of the frequency ratio; mathematically,

$$\text{Loss at } f_{Hi} = \text{Loss at } f_{Lo} \sqrt{\frac{f_{Hi}}{f_{Lo}}}$$

Example   The loss at 200 MHz for RG6/U is 3.1 dB/100 ft. To find the loss at 400 MHz, take the ratio 400/200 MHz, or 2. The square root of 2 is 1.414 × 3.1 dB/100 ft = 4.38, and is rounded to 4.4 dB/100 ft.

Coaxial cables for special purposes are larger in size than those given in Table 6.5 and usually have a solid tubing or ribbed tubing outer conductor. Coaxial cable used in the CATV industry and also for local area networks (LANs) has a thin-walled aluminum tube outer conductor and a gas-injected polyethylene foam dielectric. The center conductor may be solid copper or copper-clad aluminum. Such cables are available with a covering plastic jacket or with the bare aluminum outer conductor exposed. The jacketed cable is used near the seacoast or where a corrosive atmosphere exists. Cable sizes are available with diameters in inches ranging from 0.412 to 1 in. Metric sizes are also available. Table 6.6 gives most of the important parameters for CATV/LAN cables.

Systems using coaxial cables may be simple interconnections between devices such as video cameras, video switchers, and computer workstations, to peripheral devices. Long runs of coaxial cable may need some form of equalization to compensate for the series resistances, series inductances, and shunting capacitances. Long camera to switching systems often use cable compensation equalizers built into the camera control or switching centers. Other cable systems may use RF carriers modulated with various base-band signals where FM and or AM types of modulation are often used. Cable television systems are com-

TABLE 6.6 Attenuation and Loop Resistance for Sizes of a High-Quality CATV/LAN Cable

Parameter	Cable diameter, in				
	0.500	0.625	0.750	0.875	1.000
Loop res. copper, ohms/1000ft 68	1.24		0.55		
Loop res. alum. copper ohm/1000ft	1.70	1.10	0.75	0.55	0.40
Attenuation (dB/100ft)					
TV Ch 2 55.25 MHz	0.54	0.43	0.37	0.32	0.30
TV Ch 4 67.25 MHz	0.60	0.48	0.41	0.36	0.32
TV Ch 6 83.25 MHz	0.67	0.54	0.46	0.41	0.36
100 MHz	0.73	0.60	0.50	0.45	0.40
150 MHz	0.90	0.75	0.62	0.55	0.50
TV Ch 7 175.25 MHz	1.00	0.80	0.68	0.60	0.54
TV Ch 9 187.25 MHz	1.02	0.83	0.70	0.61	0.55
TV Ch 11 199.25 MHz	1.05	0.85	0.73	0.63	0.58
240 MHz	1.16	0.95	0.80	0.70	0.63
260 MHz	1.21	1.00	0.84	0.74	0.66
300 MHz	1.32	1.08	0.90	0.80	0.72
400 MHz	1.55	1.27	1.06	0.90	0.84
500 MHz	1.76	1.43	1.20	1.05	0.95
600 MHz	1.94	1.58	1.34	1.16	1.06

mon example for computer network communications as well as broadband local area networks.

6.1.3.1 Cable television.

Cable television systems started out as simple systems used to distribute off-air television broadcast stations to subscribers who paid for such service. Antennas were placed on either a tower or on a mountaintop. The television signals were adjusted in level, filtered, and combined into one cable feeding a cascade of repeater amplifiers to the service area. Early systems consisted of a single cable with repeater amplifiers and coupling devices tapping the cable to feed the drop cable to a subscriber's house. When the signal quality was degraded by the noise and distortion contributed by the cascade of amplifiers, the service area was terminated. To increase the size of the service area and reach more subscribers, the trunk-feeder cable system was developed. The trunk cable simply consisted of a cable and cascade of amplifiers from the head end (tower and antenna) to the farthest end of the service area. At each repeater amplifier station, bridging amplifiers were placed as needed to sample the trunk signal, amplify it, and drive a separate feeder cable. Tapping devices, either two-port, four-port, or eight-port, were spliced into this feeder cable to supply service to the subscribers. The drop cable was connected to the ports on the taps to supply television signals to the subscriber's television receiver. Such a system was called a trunk feeder dedicated system, where a port was dedicated to a house. Installation of the taps in the feeder cable caused a reduction of signal level, called *insertion loss*. As the signal into the tap decreased, the tap port signal level had to be increased to supply a proper signal level to a subscriber. When the level became too low, a line extender amplifier was installed to supply another feeder section of taps. The size of a system depended mainly on total bandwidth, which in turn governed cable size and spacing between trunk amplifiers. Usually no more than three line extender amplifiers were used in cascade, since these amplifiers had to supply high signal levels to the cascade of taps. The high gain needed to produce the high signal level caused a rapid buildup of noise and distortion products.

Cable television standards and parameters. The cable television system just described essentially supplied television antenna signals to subscribers. Therefore, the term community antenna television (CATV) was coined. Most of these television signals appeared on the cable system on the same television channel as the broadcast station. Therefore, television sets were able to be used in the same manner as from a rooftop antenna. The FM radio service was also added to the cable system (88 to 108 MHz), which appears just above channel 6. Cable operators subsequently were allowed to use the cable band above the FM band and below channel 7 (175.25 MHz), which added 9 more 6-MHz television channels. These were designated A, B, C, D, E, F, G, H, and I, or 14, 15, 16, 17, 18, 19, 20, 21, and 22. Thus the total number of channels was increased from 12 to 21. Unfortunately, there were no cable-ready television sets, and the normal television sets at the time had a VHF channel selector covering television channels 2 to 13 and a UHF dial covering 14 to 83. The UHF channels covered the frequency range of 470 to 890 MHz. Notice that UHF channel 14 video carrier is 471.25 MHz and the cable channel 14 (A) is at 121.25 MHz. Obviously no television set could tune to these nine new channels and some means had to be found. A block converter was developed that converted these new channels to UHF channels, for instance 27 to 35 (548 to 602 MHz). The block converters allowed these channels to be viewed on the UHF dial. Cable operators usually added the satellite-derived premium (pay) channels as well as some of the so-called superstations in this band. Trapping filters (negative) that denied these channels were placed on the tap port of subscribers who did not want the service. Unfortunately, cable subscribers who later decided they wanted the service caused the cable operator to dispatch a technician to go to the subscriber's house, climb the pole, and remove the trap. This of course is a costly but necessary thing to do. More satellite systems carrying cable programming were developed and cable operators had to keep increasing their system's channel capacity in order to offer these services to subscribers. Present state-of-the-art systems have an upper frequency limit of 550 or 600 MHz, which consists of 60 to 70 television channels.

This type of cable system is basically a one-way system supplying signals from a central hub (head end) downstream to subscribers, often referred to as a *tree-branch* type of network architecture. Some cable operators have offered reverse, or two-way, capability to this type of network. In some cases this was a requirement to obtain the license or franchise to operate a cable system in a town, city, or community.

Two-way cable television systems. Two-way cable television service essentially split a section of the bandwidth for upstream, or reverse, and a section for the downstream, or forward, service. When only a few channels were needed for the upstream service, the subsplit reverse system was used. This type used the 5-to-30 MHz band for the reverse channels allowing four and a half 6-MHz television channels. The band between 30 and 54 MHz was reserved as a guard band needed for the diplex filters used to separate the forward (55 to 550 MHz) and reverse (5 to 30 MHz) channels. For systems requiring more reverse capability, the midsplit system was developed. This plan actually was not a real middle split as the name implies. The forward channels were in the band of 150-to-550 MHz (approximately 66 channels) and the reverse channels in the 5-to-112 MHz band (approximately 18 channels). The band of 112 to 150 MHz was used as a guard band. A high split system pushed the crossover guard band higher with the forward band at 234 to 550 MHz (approximately 52 channels) and the reverse band 5 to 174 MHz (approximately 28 channels). In practice, not many high- or midsplit systems were built and of those that were, not many reverse channels were used. Some licensing requirements imposed by a few municipalities required a whole separate cable system built for municipal and public use where a separate upstream and downstream cable system was constructed. These special systems were called *I nets* for "institutional network." For the subsplit, midsplit, and high-split cases, the reverse amplifiers and diplex filters were installed in the same amplifier housings as the forward amplifiers. The diplex filters and shielded internal compartments separated the forward and reverse bands. Since cable loss is a function of frequency, the loss for the lower-frequency reverse band was less. Therefore in some amplifier locations a simple jumper was installed along with the filter and an amplifier was not installed. Some manufacturers made lower-gain amplifiers for the reverse portion of the plant. The network topology for a cable system is shown in Fig. 6.5.

The reverse section of a cable television system was mostly used to send upstream programming from local schools and the municipal government. Usually the subsplit technique with its four television

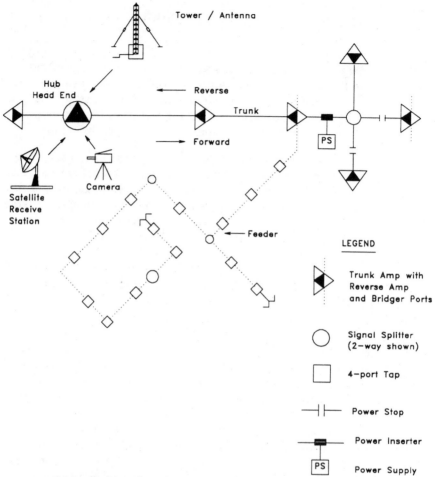

Figure 6.5 Cable television plan schematic diagram.

channels was sufficient. Cable operators found that the buildup of noise due to the reverse tree nature of the network topology caused signal degradation. To overcome this problem, a control carrier sent downstream set switches in the bridger amplifiers that disconnected all the reverse system except the path for the desired signal. Smaller cable systems with fewer branches and shorter amplifier cascades could get by without the so-called reverse bridger leg switching.

6.1.3.2 Local area networks. Local area networks (LANs) are used to interconnect computer workstations to share common equipment, such as mass storage devices, and to communicate with each other. These networks can use twisted-pair wires or coaxial cable, depending on the relative distances and the required speed of the communication methods. Various network topologies have been employed in the design of these cable systems. Essentially all the workstations have to have an interface card containing the necessary firmwave and softwave to communicate with the cable system. Such protocols are standardized by the computer industry working through the IEEE. Three principal LAN systems are Ethernet, IBM Token Ring, and MAP. Ethernet was developed by Intel Corporation, Xerox Corporation, and Digital Equipment Corporation, and conforms to IEEE 802.3 CSMA/CD standards (Carrier Sense Multiple Access with Collision Detection). The IBM Token Ring was developed by IBM Corporation and uses CSMA/CD networking conforming to IEEE 802.5 standards. MAP (manufacturing automation protocol) is also a form of token bus and conforms to IEEE 802.4 standards. A special case of MAP is called MAP/TOP (manufacturing automation protocol/technical and office protocols). All of these systems depend on the built-in software to manage and control the intercommunications among the workstations. Naturally, communication has to be bidirectional and may employ more than one cable set.

Small-size LANS.   For smaller-size systems where the cable lengths do not constitute a problem, base-band digital signals can be used. These cable systems do not contain active devices, but may contain some digitally controlled switching. Most of the problems with a small office or plant LAN of this type is a cable failure. Such failures are either a result of human error, such as cutting or driving a nail through the cable, or caused by corrosion due to moisture or chemical action. Testing for such breaks or short circuits can be done by a simple ohmmeter continuity test on the open ends of a cable section.

Coaxial cable LANS.   Most coaxial cable LAN systems use some form of modulated carrier technique. A series of carrier frequencies properly spaced through the bandwidth with the modulation type chosen to restrict the bandwidth of each carrier is often used. A common type is the CATV midsplit or high-split system with the television carrier frequencies NTSC AM VSB modulated. If FM, AM, or FSK (frequency shift keying) is employed as the modulation of these carrier frequencies, this system could be used as a bidirectional multichannel LAN. Some MAP/TOP systems are of this type and are used to control robotic devices in the automotive manufacturing industry.

It was mentioned previously that the tree-branch forward architecture operating in the reverse mode causes the buildup of noise. The forward system contains all channels (RF carriers) of different fre-

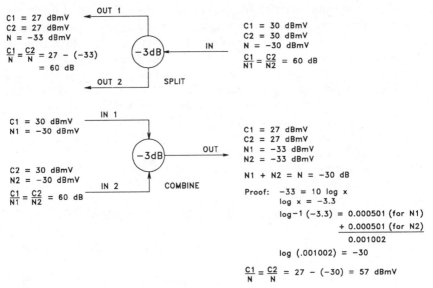

Figure 6.6   Carrier, noise splitting, and combining.

quencies combined on a single cable from the hub or central point downstream to all the branches. The reverse system picks up each frequency from a different branch and the signal is combined only at the hub point. A simple demonstration of this effect, using a single hybrid splitter-combiner device, is shown in Fig. 6.6. Notice in the Split case that the output C/N is the same as the input, whereas in the combine case the C/N is deteriorated 3 dB. Naturally, the more combining points, the more noise funnelling occurs, thus reducing the C/N. This phenomenon has to be taken into account during system design in order to preserve acceptable reverse C/N values.

Another problem that constantly plagues maintenance personnel is that of maintaining tight connectors. Loose connectors cause impedance mismatches, which in turn cause signal reflections, resulting in signal-level variations and echoes. Also, signal leakage noise ingress results, causing more problems. As cable television people know, the FCC has strict guidelines, procedures, and recordkeeping requirements of plant signal leakage control in the aircraft and air safety bands of the frequency spectrum. Plant shutdown and fines are some of the penalties. Therefore, a tight cable plant will keep leakage to a minimum and preserve signal quality.

6.2   Fiber-Optic Cables

As bandwidth requirements increased for coaxial cable, the size of cable and the gain of amplifying devices became greater. The development of fiber-optic cable and optical devices was done in a timely manner. Fiber-optic cable consists of one or more glass fibers clad with a higher-refractive-index coating than the main glass fiber. A Kevlar jacket provides mechanical strength to the fiber and protects the fiber from knicking or scratching. The fibers are then placed in another jacket called a *buffer tube*. Often there will be two, three, four, five, or six clad jacketed fibers in a color-coded buffer tube. For instance, 18-fiber cable may have three buffer tubes (brown, orange, green), containing six jacketed (brown, orange, green, blue, white, and slate) fibers in each of the three buffer tubes. Fiberglass strands, a stronger fiberglass strength member, plus a good polyethylene jacket give the needed strength to the cable to allow pulling from a reel, for utility-pole placement, or for pulling through underground conduits during plant construction phases. An example of fiber-optic cable construction is shown in Fig. 6.7.

In practice it has been demonstrated that present-day fiber-optic cable is more rugged than coaxial cable as far as resistance to breakage from pulling, crushing, and bending. However, splicing does present a problem. Essentially the cable has to be stripped back to expose the glass fibers, which then are cleaved to a specific angle or flush cut and

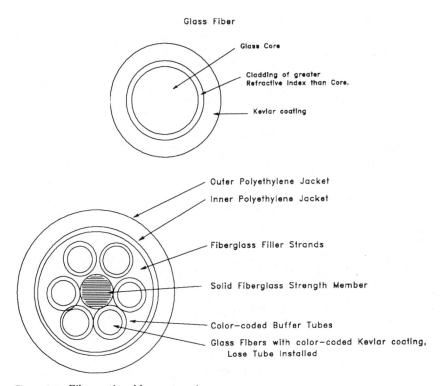

Figure 6.7 Fiber-optic cable construction.

then fusion-spliced or mechanically spliced. Fusion splicing generally provides a lower-loss splice on the order of 0.01 to 0.05 dB, while a mechanical splice from about 0.08 to 0.2 dB. A mechanical splice has the capability of being disconnected so as to insert an optical instrument. Both types of splices are placed in a slotted metal tray with extra glass fiber stored in the tray in a figure-eight pattern. The whole tray is mounted in a specially constructed plastic enclosure the size of an automobile muffler. This splice, if pole-mounted, is usually placed approximately 25 ft from a pole and extra fiber-optic cable is hung on both sides in a flat looped fashion. The fusion-splicing equipment is quite expensive and a high degree of expertise is needed by the splicing personnel. Many splicers have stated that cleaving and cleaning the glass fiber is the most important part of the operation.

The transmitting and receiving end of a fiber-optic cable system requires an intensity-modulated laser as the transmitter and a photodetector diode receiver. The modulated laser can be of the lower-cost Fabry-Perot type or the more spectrally pure distributed feedback (DFB) type. The receiver is simply a connection of the optical fiber to the sensitive face of a photodetector diode whose output is an electrical signal proportional to the modulation. The bandwidth of the modulating signal is extremely large and is actually more limited by the modulating and receiver electronic circuits than the optical fiber. Naturally, the laser as well as the receiver can be overdriven, with the usual resulting signal distortion. Therefore, a properly designed fiber-optic communication system will not allow overmodulation of the laser output power which drives the glass fiber. At the receiver an optical attenuator may be needed so as to not overdrive the receiver. Typical laser outputs are in the range of -3 to $+2$ dBm of optical power. Receiver thresholds require a minimum input of from -25 to -18 dBm. The typical loss of optical fiber is in the range of 0.25 to 0.8 dB/km. Therefore, 18 dB of cable loss results in a distance of 18/0.25 = 72 km, or 45 miles for 0.25 dB/km cable. The long-haul communication possibilities should be obvious. If 6-km reels of cable are available, then 12 reels will be needed, which results in five splice locations, not counting both ends. If loss of 0.02 dB per splice is obtainable, total splice loss is 0.1 dB. Adding this to the 18-dB cable loss give a total path loss of 18.1 dB. If a receiver with a -20-dB threshold and a laser with 1 dBm of output is used, a 2.9-dB margin results, as shown in Fig. 6.8. The foregoing illustration is very minimal and a more acceptable signal margin would be somewhere between 6 and 9 dB. This allows extra splices and connections to be installed in the fiber run and preserves signal quality.

The laser operating frequencies are given as wavelengths and three main operating windows appear for present-day glass fiber. They fall

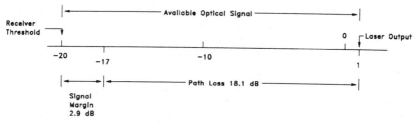

Figure 6.8 Optical system signal analysis.

in the ranges of 850, 1300 to 1310, and 1550 to 1555 nanometers (nm,10^{-9} m) of wavelength. These frequency bands do not fall in the area of visible light and hence cannot be seen. However, they can cause damage to the eye and therefore caution should be observed and one should not look into any laser output connectors. Suitable warning labels are used and many companies require employees to wear protective glasses.

6.2.1 Fiber-optic cable systems

Probably one of the first users of fiber-optic systems was the telephone industry. Since modern long-haul telephone systems use high-speed digital transmission techniques that require more bandwidth than the twisted copper wires, fiber-optic cable allows thousands of digitally encoded telephone channels per fiber. The cable television industry uses fiber-optic systems in transporting the whole CATV band deep into the cable system and thus improving their signal quality by reducing the length of long cascades. Such a technique is simply illustrated in Fig. 6.9. Cable television operators have also used fiber-optic

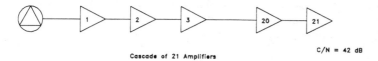

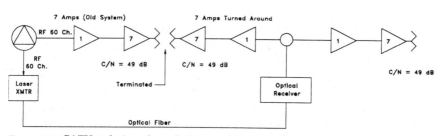

Figure 6.9 CATV technique for reducing amplifier cascades.

techniques to transport satellite received signals from remote antenna sites to the hub location for processing and distribution to subscribers. Television broadcast stations are also looking at fiber-optic techniques as studio to transmitter links, as well as to connect remote facilities together.

LANs are using fiber-optic methods to connect computer workstations together as well as to the file-server mass storage library. LANs use more actual fibers than cable television simply because LANs typically have to be multichannel bidirectional systems. The optical splitter has been developed, allowing optical cables to branch off to other areas, thus enhancing the overall network topology possibilities in LAN design. Optical splitters and dividers can be made in a variety of optical loss/ports values and can have typically two, three, or four ports.

6.3   Cable System Testing

Testing of both cable and fiber-optic communications systems requires a variety of sophisticated instruments and procedures. Testing is performed for two main reasons, one is to test to prove performance specifications, the second is during troubleshooting and maintenance. Usually at the end of the system construction phase, exhaustive testing is performed to determine if the system meets specifications. These test results should be retained as a zero-level base used to compare later test results which will indicate whether system performance is either stable or deteriorating. Cable communications systems requiring a high degree of reliability should retain an on-site trained maintenance staff with proper test equipment and spare parts. In many cases where the ultimate in reliability is not demanded, a maintenance or service company can be engaged to make periodic testing, maintenance, and cleaning visits to the facility with emergency service available.

6.3.1   Coaxial cable testing

Coaxial cable system testing for a passive cable system, consisting of just cable and no active devices, is simply a test of cable and connectors. Such a system can be used to connect computer or television equipment to various devices. Operation of this type of system depends on the integrity of the cable. Since coaxial cable has a distributed capacitance, inductance, and resistance per unit length, testing for these parameters will prove that the cable is operational. The open- and short-circuit condition can be tested by a simple ohmmeter continuity test. The distance to a fault can be made using a time do-

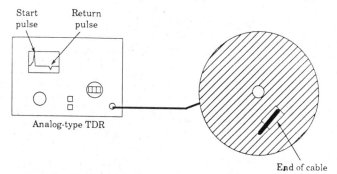

Start pulse   Return pulse

Analog-type TDR

End of cable

Figure 6.10 Testing of cable length on reel using the TDR.

main reflectometer (TDR), which transmits an electrical pulse on the line and measures the presence of any reflection. A reflection will indicate a short-circuit or open-circuit condition. Some TDR instruments have an oscilloscope readout (Tektronix models 1502, 1503) where the polarity of the reflected pulse indicates whether the fault is open- or short-circuited. The distance to the fault is calculated by the instrument by measuring the time between the transmitted and received pulses (round-trip time) and dividing by 2. The Tektronix 1502 with the oscilloscope readout will tell the distance to the fault on the distance dials as well as the nature of the faults along the line. An optional recorder can be connected to the instrument which will produce a chart of fault or faults versus distance for a permanent record. This procedure is illustrated in Fig. 6.10 where the instrument is being used to measure the amount of cable on a reel. One end of the cable is left open to cause the TDR to read the open-circuit fault at the end of the reel. The distance measured is the length of the cable on the reel. Some TDR instruments indicate the distance to the fault with a numerical liquid crystal display (LCD) readout and a light emitting diode (LED) indicating the open- or short-circuit condition.

If the length of the cable run is known and the capacitance and inductance per foot are known, then by using an LCR meter the distance to a fault can be calculated. A short-circuited cable allows the inductance to be measured and an open-circuited cable is mostly capacitive. An example of finding the distance to a fault using this technique is shown in Fig. 6.11.

6.3.1.1 Coaxial cable operational tests. Testing on cable systems using active devices means testing the whole system for operational parameters. If the cable is suspected of not allowing the system to operate, tests on the cable for continuity and breaks can be done using the testing method just discussed. Such cable systems using active devices in-

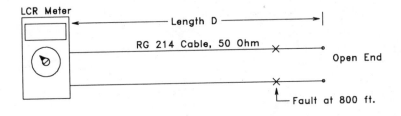

Calculations

In a vaccuum, $\nu_p = 980 \times 10^{-6}$ ft/sec.

In cable, $\nu_p = 980 \times 10^{-6} \times 0.66$
$= 649 \times 10^{-6}$ ft/sec

$\nu_p \approx \dfrac{1}{\sqrt{LC}}$ and $Z_o \approx \sqrt{\dfrac{L}{C}}$

D = 1000 ft
ν_p = Velocity of Propagation = 66%
C = 31 pf/ft
Zo (nominal) = 50 Ohms

L calculates to 0.0765 μH/ft.

If fault is a short circuit, LCR Meter will indicate L = 61.2 μH.
If fault is an open circuit, LCR meter will indicate C = 0.0241 μf.
If LCR Meter indicates 0.031 μf no fault found for 1000 ft of open end cable.

Figure 6.11 Cable testing with LCR meter.

volve carrier frequencies modulated with various signals. Cable television systems and some LAN systems are of this type. The network topology may be of the tree-branch type with reverse-channel capability or a redundant ring type with optional routing to distribution points called nodes. A study of the network topology is necessary in order to establish a test plan that is able to prove performance and identify problem areas. Since cable television systems constitute the greatest number of users, they will be used as examples for most cases.

Frequency response tests.   Frequency response of cable television systems is a most important parameter. Cable losses increase with increasing frequency. These losses are caused by the distributed inductance and capcitance inherent in the cable's construction. The cable television industry uses what is known as the unity-gain building block for cable design, as illustrated in Fig. 6.12. In order to establish proper signal input to the amplifying integrated circuit (IC), an equal-

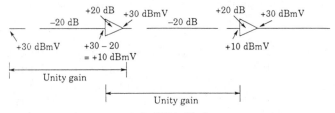

Figure 6.12 The unity-gain building block.

izer is needed to compensate for the cable loss at the higher frequencies. Also, the output of the repeating amplifier is tilted upward a specified amount so that full gain is used at the upper frequency limit. According to theory and as proven in practice, this limits distortion and noise to practical values. Cable size, amplifier gain, noise, and distortion control the practical distances covered by a cable system. Proper control of signal level throughout the frequency band is maintained by a testing and maintenance program. Tight, leak-free connections to the various splitters, couplers, amplifier housings, and power-supply devices keep the signal inside the cable plant and noise outside. Also, loose connectors cause cable-plant impedance fluctuations at various frequencies, which cause signal-level variations and ghosting (echoes).

Frequency-response testing of a cable plant involves the use of a specially designed signal sweep system. In the past, maintenance personnel merely caused a signal carrier of constant amplitude to sweep through the frequency band from the head end or hub location. At the output or input test port at any amplifier location a sweep receiver could monitor the action of this carrier. A special control carrier placed below the low-frequency limit carried the control signal and told the receiver to start measuring. This signal sweeping through the system bandwidth caused interference with the system's signals and hence the picture quality of the cable television service. The sweep speed and signal level were controlled to minimize this problem. However, as far as performance was concerned, this sweep system provided an accurate test of the system frequency response.

In use today are two sweep-testing products that eliminate all or nearly all of the interference to the television pictures. The instrument that eliminates all of the interference is the Tektronix CATV Sweep System. This instrument consists of two devices, a sweep transmitter (Tek model 2721) and the receiver (Tek model 2722). The transmitter has a frequency-selectable control carrier that may be set anywhere in the test band (5 to 600 MHz). The receiver is battery-operated and has a hook bracket for hanging on the cable-supporting messenger strand. The way this system works is the sweep carrier of proper level is positioned on the video carrier frequency but not activated. During the vertical interval of the video signal, the sweep carrier is injected, thus the presence of the sweep signal is not seen on a subscriber's TV set. Each TV channel is swept through the band. Where no carrier is present, the sweep signal is activated. The receiver is programmed to decode the control carrier and to find, display, and store the sweep signal information. The receiver displays the sweep response on an LCD screen or optional chart recorder and has a built-in random access memory (RAM) sufficient to store 50 traces of

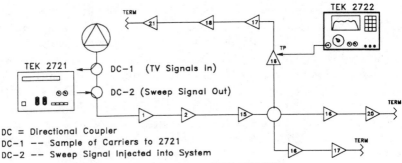

DC = Directional Coupler
DC-1 -- Sample of Carriers to 2721
DC-2 -- Sweep Signal Injected into System

Figure 6.13   Automatic sweep system using TEK 2721/2722.

sweep information. Since this equipment is completely programmable, its use on both subsplit, midsplit, and high-split systems for both the forward and return path are possible. Optional software for an IBM-compatible personal computer (PC) is available to process the stored trace data for future display or analysis. A programmable keyboard on the TEK-2722 receiver allows alphanumerical identification of the stored and displayed sweep traces for selecting, measuring, and displaying individual signal carriers or the whole spectrum of carriers. An ac-dc voltmeter and a temperature probe are also included. Use of this instrument is shown in Fig. 6.13. A full battery charge on the receiver can give about 8 hours of use, which amounts to a full day's work for a sweep technician.

Use of a signal-level meter is important in setting up the amplifier cascades for a cable television system. Several types are available from a number of manufacturers. Some are tuned with a knob and dial and some are tuned on a keypad with tuning at every tenth of a megahertz, or on an assigned standard television channel video frequency, or the cable television harmonically related carrier (HRC) system. The measured signal level is displayed on a tautband electromechanical meter movement or LCD alphanumeric display. The selected frequency may be displayed on a dial on some meters or on an LCD display on others. Prices range from about $600 to $3000 for instruments that can be tuned from up to 600 MHz. The proper way to activate a cable system after construction and before subscribers are connected is to activate the pilots or control carriers and the lowest and highest carriers at the bandwidth limits and then set them to the correct signal levels. Each amplifier in the system is then individually adjusted in signal level starting with the first amplifier in each cascade and progressing to the end. This procedure is called turn-on and rough balance. Then the leakage-detection carrier, if not already activated, is level set and added to the system for leakage testing. When the system is made leak-free by tightening loose connectors, devices, and equipment housings, then final balancing and adjusting should take place. As mentioned before, loose connections cause impedance mismatches, which cause reflections (ghosting) and signal-level variations. It would be wrong to attempt final balance on a system that is not tight. Compensating for a poorly constructed system, that is, with loose connections by the final balance, will almost guarantee a poor cable system. Final balance uses a sweep system to measure the system response, which indicates the areas of adjustment such as amplifier gain, slope, equalization, and in-band response. The unity-gain building block approach calls for each amplifier in the trunk cable cascade to have the same input signal level and the same output level and consequently the same gain. Manufacturers try to make their amplifiers with identical specifications and most of them do a good job. However, each amplifier has its own signature and the effects of any midband slump in frequency response can accumulate through the cascade. Therefore, the adjustment of, for instance, every fifth amplifier may be needed to correct for this effect. Amplifier manufacturers usually have available plug-in interstage trimming networks where response corrections are made by selecting the proper network circuit board and setting any of the variable control elements. For two-way systems, the reverse-channel balance has to be done on each upstream cascade separately. Unfortunately, the sweep transmitter injection point might be on a pole, necessitating a vehicle and connecting lead to be placed at the pole the entire time the cascade is balanced. It should be evident that for both the forward and reverse balance, all amplifier stations have to be visited twice. One method used to solve this problem is a reverse-sweep transmission system that is battery-operated and injects a sweep signal upstream. A sweep receiver with a CRT display is viewed by a television camera that sends a television picture on a downstream channel to be viewed on a small battery-operated television set at the sweep injection site. This method is shown in Fig. 6.14. The reverse signal for a subsplit system of four channels can be generated by a comb generator and the receiver can be a spectrum analyzer. Forward and reverse amplifier alignment adjustments can be made simultaneously at each amplifier location. Data for the forward path can be recorded by the Tektronix Sweep System and for the reverse path on the television receiver VCR.

The signal-level meter presently used by cable television systems and coaxial cable LAN systems is a highly developed result of the earlier field-strength meters. The field-strength meter was a frequency-selectable voltmeter with, in many cases, a built-in calibrated loop antenna. The meter scale was calibrated in microvolts per meter, the standard unit of field strength. Present-day meters are still frequency-

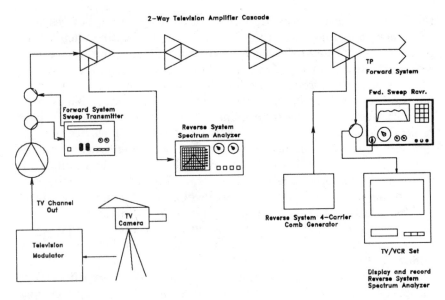

Figure 6.14 Alignment of both foreward and return paths for a cable television section of plant.

selective voltmeters with a 75-ohm input impedance, a precision step-input attenuator and are calibrated in microvolts or dBmV. The dBmV in decibels referred to a level of 1 mV across a 75-ohm resistor (0 dBmV). Signal levels above (+) and below (−) this level are read on the meter face as + or − the indicated dBmV. Meters presently available have an added feature that extends the usability of the instrument. Measurement of noise level, hum, and low-frequency disturbance can be made using most present-day meters. Digital (keypad) selection of frequency, LCD display of frequency and/or signal level, as well as digital (RAM) storage of data are some of the other features. Some manufacturers provide a built-in sweep method of frequency-tuning the instrument, and by having the start sweep signal available to drive an oscilloscope, a simple form of spectrum analyzer extends the use of the instrument. One manufacturer uses a large-screen LCD and presents the frequency swept-signal levels as a histogram display simulating a spectrum analyzer. The added capability of selecting and displaying one television channel's video and audio carrier allows portions of the overall spectrum to be measured channel by channel. The latest offering by Comsonics Corporation, the manufacturer of this type of signal-level instrument, is a handheld meter controlled by a keypad and displaying the channel signal-level measurements on an LCD screen. Frequencies can be selected using the television channel number on the keypad or by frequency value by pressing the fre-

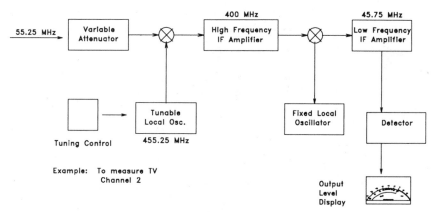

Figure 6.15   Double-heterodyne signal-level meter.

quency directly on the keypad. Thus this instrument, as well as just about all the signal-level instruments, can be used for LAN systems where carrier frequencies are not necessarily the same as television carriers. The accuracy of signal-level measurements for many signal-level instruments in use today ranges from ±0.25 dBmV to ±1 dBmV. The less-accurate instruments are less expensive and are used by installation and service personnel. Spectrum analyzers can usually provide accuracies of nearly the same value. However, the operator has to view a screen and hence may introduce more operator error. Most signal-level meters contain circuitry similar to that of the spectrum analyzer and are usually of the double-superheterodyne type. Figure 6.15 is block diagram showing this method. The signal-level meter is the most-often-used instrument for cable television systems and coaxial cable LAN systems employing RF data carriers. Most instruments are rugged and weather-protected for field use.

Signal distortion tests.   Signal distortion causes picture impairments, defects in cable television systems, and the bit error rate (BER), causing data errors in coaxial cable LAN systems. The principal causes of signal distortion are incorrect signal levels overdriving the repeater amplifiers or simply a defective amplifier. Signal distortion caused by amplifiers is usually some form of harmonic distortion, usually either second- or third-order distortion. Higher-order distortions are usually too small to cause problems or occur out of the signal bandwidth. A study of amplifier cascade theory given in several of the references at the end of this chapter is advisable. However, the old rule of trading signal level for bandwidth is still good. For example, an amplifier carrying just a few signals can provide extremely high output level with minimal distortion. If 60 television carriers covering the bandwidth of

Cascade Noise Figure (NF) = NF(single amp) + 10 log N
Cascade Carrier-to-Noise Ratio (C/N) = C/N(single amp) − 10 log N
Cascade 2nd Order Distortion = 2nd Order(single amp) + 15 log N
Cascade Single Triple Beat (STB) = STB(single amp) + 20 log N
Cascade Composite Triple Beat (CTB) = CTB(single amp) + 20 log N
Cascade Cross-Modulation (XMOD) = XMOD (single amp) + 20 log N
Cascade Hum Modulation = Hum-Mod(single amp) + 20 log N

N = Number of amps in cascade
The values for a single amp are given in the manufacturer's specifications.

Figure 6.16   Cascade buildup of noise and signal distortion.

50 to 450 MHz are applied to the amplifier, each carrier will have to be reduced far below the level of the case where only a few signals were applied. Essentially the amplifier's power handling capabilities have to be expanded over many carriers instead of just a few. Cascaded amplifier theory has shown that second-order distortion buildup is partly on a power and partly on a voltage basis, hence $(15 \log N)$, where N is the number of amplifiers in a cascade. Third-order distortions are divided into single triple beats, cross modulation, and composite triple beat. Most forms of triple beats are called *intermodulation*. Buildup of third-order distortion is on a voltage basis, hence $(20 \log N)$. Figure 6.16 gives the cascade mathematical relationships for signal distortions, noise, and low-frequency disturbances (hum). The single-amplifier specifications are found on manufacturer's data sheets. The cascade buildup assumes that each amplifier has identical characteristics and that the input signal levels are the same. Most manufacturers do indeed have the ability to provide amplifiers that are close to identical. However, the effects of any slight nonlinear characteristic can accumulate in the cascade causing poor performance at the cascade ends. That is why final signal sweeping and balancing are done before full turn-on and system activation. Good maintenance practice should include a sweep testing program once a year minimum, and any problem areas corrected before the situation becomes epidemic.

Composite triple-beat distortion is the measurement of choice of most cable television personnel relevant. This distortion test requires the system to be fully loaded with the television carriers and that the carriers have the modulation removed. There are two ways to do this depending on the type of equipment at the head end. If heterodyne processors are used they can be put in the standby carrier mode, and for modulators the modulation connector can be removed. Extra modulators of the tunable variety can be used to activate the rest of the channels needed for the full-load case. The second method uses a multiple carrier generator which essentially provides all carriers (unmodulated) to drive the system. In the case of a 50- to 450-MHz system which requires 60 channels to become a fully loaded system, the 60-channel generator is quite large and expensive. To make this measurement with all unmodulated carriers present driving the system, a bandpass filter and spectrum analyzer are placed at the longest amplifier cascade location. A table of the buildup carrier beats for various system channel capacity is given in Table 6.7. It should be seen from the table that the area of the greatest number of possible beats is around channels 32 and 33 (271.250 and 277.250 MHz) for a 60-channel system with an upper frequency limit of 450 MHz. Therefore, the bandpass filter should be 4 to 6 MHz wide and centered on channel 32 or 33. Now, the technique is to read the signal level of channel 32 or 33 at the cascade and then turn off the carrier at the head end. The spectrum analyzer will show buildup of beats in the vacant area of the channel 32 or 33 band. The gain of the spectrum analyzer may have to be increased in order to make a more accurate measurement if the beat level is quite low. The bandpass filter protects the analyzer from being overdriven by the other carriers when the gain is increased. This measurement technique is shown in Fig. 6.17. The spectrum analyzer should be tuned to the center of the bandpass filter with a frequency span per division on the order of 100 kHz and a 30-kHz resolution bandwidth (RBW).

To make a second-order single-beat measurement, two carrier frequencies should be chosen so that their sum or difference frequency will fall close to a standard channel. For instance, for a 300-MHz 35-channel system, the carriers at channel W (35) at 295.25 MHz and channel M (26) at 235.25 MHz will produce a beat (60 MHz) which falls between channels 2 and 3. The procedure is to place these two frequencies on the system at normal level with the pilot carriers (both low and high pilots) on the system as well and at proper level. At the last amplifier a spectrum analyzer and a bandpass filter are stationed to make the measurement. The bandpass filter should be able to pass the 60-MHz signal and have a bandwidth of 4 to 6 MHz. A test carrier should be activated near 60 MHz and the level measured by the analyzer. Then this carrier should be turned off and the analyzer gain increased enough to see and measure the beat produced by the difference of the test carriers. If, however, the beat appears too small to be measurable, the level of the test carriers can be increased 10 dB, the pilot carriers remaining at their normal levels. Now the analyzer can probably make the measurement. In the first case the difference between the test carrier level (near 60 MHz) and the beat level as measured by the spectrum analyzer in decibels is the value. In the second case with the signals increased 10 dB, the measurement will actually be 10 dB better. The measurement method for second-order single-

TABLE 6.7   System Beats

Channel No.	35 Channels	40 Channels	52 Channels	60 Channels	77 Channels	36 Channels 234–450 MHz	53 Channels 234–550 MHz
2	159	235	435	615	1104		
3	171	240	456	640	1137		
4	180	251	473	661	1167		
5	31	36	48	56	73		
6	31	36	48	56	73		
7	342	458	788	1048	1707		
8	345	464	800	1064	1731		
9	348	469	811	1079	1755		
10	349	473	821	1093	1777		
11	350	476	830	1106	1799		
12	349	478	838	1118	1819		
13	348	479	845	1129	1839		
14	274	348	644	868	1450		
15	288	384	666	894	1485		
16	299	398	686	918	1517		
17	308	409	703	939	1547		
18	316	420	720	960	1576		
19	323	429	735	979	1604		
20	329	438	750	998	1631		
21	334	445	763	1015	1657		
22	338	452	776	1032	1682		
23	345	479	851	1139	1857		
24	342	478	856	1148	1875		
25	337	476	860	1156	1891		
26	332	473	863	1163	1907	306	676
27	326	469	865	1169	1921	323	701
28	321	464	866	1174	1935	339	726
29	315	458	866	1178	1947	354	749
30	309	451	865	1181	1959	368	772
31	301	443	863	1183	1969	381	793
32	293	435	860	1184	1979	393	814
33	283	427	856	1184	1987	404	833
34	273	419	851	1183	1995	414	852
35	260	410	845	1181	2001	423	869
36	245	400	838	1178	2007	431	886
37		389	830	1174	2011	438	901
38		377	821	1169	2015	444	916
39		364	811	1163	2017	449	929
40		349	800	1156	2019	453	942
41		331	790	1148	2019	456	953
42			775	1139	2019	458	964
43			761	1129	2017	459	973
44			747	1118	2015	459	982
45			733	1106	2011	458	989
46			719	1093	2007	456	996
47			704	1079	2001	453	1001
48			688	1064	1995	449	1006
49			671	1048	1987	444	1009
50			653	1031	1979	438	1012

TABLE 6.7   System Beats (*Continued*)

Channel No.	35 Channels	40 Channels	52 Channels	60 Channels	77 Channels	36 Channels 234–450 MHz	53 Channels 234–550 MHz
51			634	1013	1969	431	1013
52			613	995	1959	423	1014
53			589	977	1947	414	1013
62				959	1935	404	1012
63				940	1921	393	1009
64				920	1907	381	1006
65				899	1891	368	1001
66				877	1875	354	996
67				854	1857	339	989

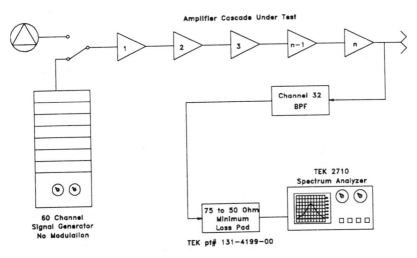

Figure 6.17   Composite triple-beat measurement.

beat is shown in Fig. 6.18. Often this test uses three carriers for third-order distortion testing on LAN coaxial cable systems. For the reverse case, the 60-channel selectable generator will be at the end of the cascade and the analyzer and filter placed in the head end. Usually a vehicle with commercial power available will be necessary at the end of cascade site. The methods are essentially the same but the selection of carriers is different. For the case of subsplit reverse with only four television carriers available, distortion is not a problem but noise buildup is the limiting factor. In general, when all carriers are modulated, the beats produced by sum and difference frequencies falling near or on other carriers cause the modulation products to mix, caus-

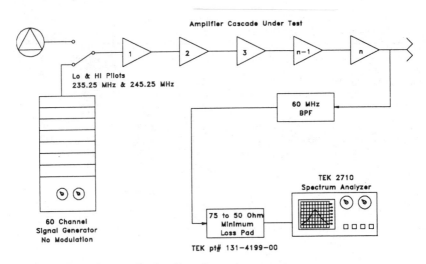

Figure 6.18 Single second-order distortion measurement.

ing picture impairment. Since each carrier has its own modulation, the effects of any cross-modulation appear as a windshield wiper effect occurring in a random fashion. To check for this type of problem, the test signal generator is modulated from a common generator, usually a squarewave generator operating at about the television horizontal frequency of 15,735 Hz. Now the cross-modulation effect will "stand still," and a meaningful measurement can be made. The technique is to modulate all carriers synchronously with the same modulating signal, and a carrier at about channel 32 or 33 will have its modulation removed. The difference in modulation readings before the modulation is removed and the value after the modulation is removed is the measurement value. A spectrum analyzer can be used as a receiver or a high-quality signal-level meter with a video output option will also do the job. This measurement is shown in Fig. 6.19. The fact remains that when third-order distortion becomes a problem, then both cross-modulation effects and composite triple-beat effects both occur. Therefore, in many instances the method chosen depends on the availability of equipment.

Noise testing, carrier-to-noise ratio.   Noise is always a problem with any communications system. Cable systems have to live with the thermal noise floor of −59 dBmV at 68°F (20°C). Loose connections allow the signal to leak out and noise to enter. The effect noise has on television pictures is snow and confetti falling over the images and is very annoying. The cable television industry has several instruments and methods available to measure noise, with the principal measurement defined as carrier level to noise level (C/N) ratio. When the carrier

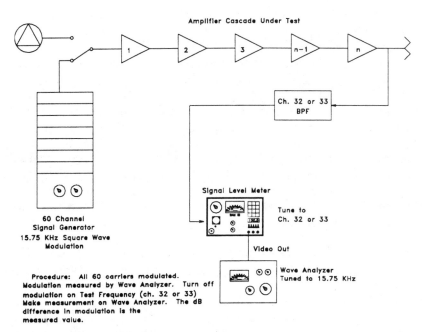

Amplifier Cascade Under Test

60 Channel
Signal Generator
15.75 KHz Square Wave
Modulation

Ch. 32 or 33
BPF

Signal Level Meter

Tune to
Ch. 32 or 33

Video Out

Wave Analyzer
Tuned to 15.75 KHz

Procedure: All 60 carriers modulated.
Modulation measured by Wave Analyzer. Turn off
modulation on Test Frequency (ch. 32 or 33)
Make measurement on Wave Analyzer. The dB
difference in modulation is the
measured value.

Figure 6.19   Cross-modulation measurement.

level is read in dBmV and the noise level is also read in dBmV, then the C/N ratio is the dBmV difference. Several manufacturers of signal-level meters have C/N measurement features. A spectrum analyzer such as the TEK 2710 can make C/N measurements. This test is simply made usually at the cascade end for a worst-case condition. The procedure using a signal-level meter with the C/N test feature is to read the signal level near a place in the spectrum with a couple of vacant channels, and then to tune to the middle of the vacant area, activate the switch for the C/N test, and remove enough attenuation to allow a reading of the noise level. Activating the switch widens the video bandwidth to the noise-specified value of 4 MHz. The signal level should be at least +20 dBmV for most instruments to give a valid reading. For example, if the signal level is read adjacent to the vacant area at +28 dBmV and the noise was read as −20 dBmV, then the C/N is 48 dB. Most meters are able to read down to about −30 dBmV. It may be necessary to go to a bridger test point to obtain sufficient signal level.

Hum and low-frequency noise.   Hum and low-frequency disturbances are usually related to power supply problems. Since 60-V, 60-Hz alternating current is carried on the cable to power the amplifiers, the amplifier power supply has to have its direct current pure and free from any ac ripple which could cause modulation of the video signal if it becomes great enough. To measure hum modulation, which is measured as a percent of total modulation, sounds complicated. However, most high-quality signal-level meters have a hum test function. To perform this test, the signal-level meter is connected to a test port of the amplifier or device being tested and tuned to an unmodulated carrier. The signal-level meter is switched to the hum test function and the percent of hum modulation is simply read on the meter scale. The video output connection on the signal-level meter can be connected to an oscilloscope where the full modulation will be read as 100 percent; then the screen can be calibrated in percent per division, and then when the unmodulated carrier is tuned the reading of hum modulation can be read and the calculation in percent made.

FCC system leakage and test methods. Since tightness of connections, amplifier housings, splitters, and couplers are so important to the overall operation of a cable system, a test of signal leakage is important. Also, it is now mandatory for any cable system, including LANs that carry signals in the aircraft communication and navigation bands to test the system for leakage on a quarterly basis with one covering 100 percent of system. The frequency bands where radiation or system leakage becomes a problem are 108 to 137 MHz (above the FM band) and 225 to 400 MHz. Also, cable systems have to offset their carrier frequencies so that they will not fall on any of the aircraft frequencies if some leakage did occur. These offsets require a 12.5-kHz ± 5-kHz shift either above the normal carrier frequency, called a *positive offset,* or below the normal carrier, called a *negative offset,* for carriers in the 118 to 137-MHz band and the 225 to 328.6-MHz band. A 25-kHz ± 5-kHz offset either above or below the normal carrier assignment is required for frequencies in the bands of 108 to 118 MHz and 328.6 to 335.4 MHz. Accurate frequency measurements are now needed and required to prove compliance with the offset requirements. Simple measurement of the video carrier frequency for a television modulator is done by measuring the RF output with no modulation with the sound carrier level reduced or turned off. The frequency counter should be recently calibrated and accuracy proved before performing measurements used to prove compliance with FCC cable television regulations. This procedure was discussed in Chapter 4, under "Video and Audio Carrier Frequency Tests." To test the sound carrier, the video carrier has to be filtered out or turned off, and the level of the unmodulated sound carrier increased sufficiently for the frequency counter to make the measurement. This procedure is an out-of-service test and a spare modulator can be substituted during the testing. Frequency testing of a television channel without removing it from service is possible using one of several methods. Manufacturers of cable television test equipment have two instruments that will do the job. One manufacturer makes what is known as a signal-stripping amplifier. The channel to be tested is selected, fine-tuned by the instrument, amplified, and stripped of video modulation. Thus the clean video carrier is presented and measured by an accurate frequency counter. A special feature of this instrument is that a precise intercarrier (4.5-MHz audio subcarrier) is also presented to the counter for measurement. Another manufacturer has essentially the same device integrated into a frequency counter of sufficient accuracy. This single-unit system is called a *tuned frequency counter* and can also be used as a frequency counter for other purposes. For a television channel on a cable system the FCC specifies the video carrier offset accuracy and the audio intercarrier accuracy. For example, + offset television channel 14, which requires a 12.5-kHz offset, should be measured as 121.250 MHz + 12.5 kHz is 121.2625 MHz ± 5 kHz and the intercarrier at 4.5 MHz ± 1 kHz (4.501 or 4.499 MHz).

Another method of frequency testing without removing the channel from service is the substitution method. Essentially a stable selectable signal generator is frequency-matched by detecting the match in a receiver and then measuring the matched generator frequency with a counter. It is important that the levels of the carriers to be measured are high enough for the counter to make an accurate reading. These methods of frequency measurement are shown in Fig. 6.20.

The method of leakage detection for cable television systems and coaxial cable LANs using carrier frequencies in the aircraft bands has been spelled out very well by the FCC in parts 76-611 and 76-614 of their regulations. Historically, plant leakage was found to be a problem when airline pilots reported cable television leakage causing interference with communications and navigational instruments. It was feared this could cause air safety problems. The FCC then passed rules and procedures requiring cable operators to maintain a so-called clean plant. Essentially a 5-year period was given to accomplish this and interim procedures were enforced. It is required that a continuous monitoring and plant correction program be employed, which requires about three-fourths of the plant be tested every 3 months, and once a year a ground-base measurement of all leaks made and a figure of merit called the *cumulative leak index* be calculated. This information has to be filed with the FCC on Form 320 by every July 1st. The FCC has stipulated the method for measuring a leak. The leak is measured using a tuned dipole placed 10 ft (3 m) longitudinally with the cable with the leakage read on a signal-level meter. The basics of this procedure are shown in Fig. 6.21. Although the FCC does not specify the test frequency, essentially leakage at any frequency in the aircraft

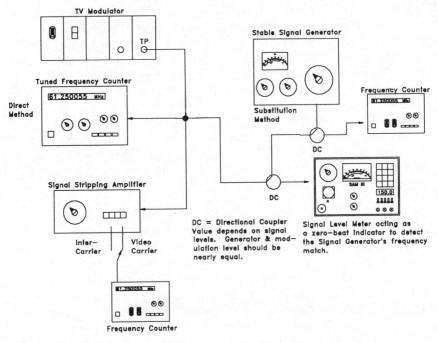

Figure 6.20   Television channel frequency measurement.

bands is meant. Cable operators choose a frequency close to the aircraft bands, usually around 108 MHz. Higher frequencies tend to cause leaks of smaller levels. Therefore, the lower frequency is a good choice such as 108.625 MHz which is a worst-case condition. At this frequency the signal-level meter (SLM) reading for the maximum allowable signal strength of 20 μV/m is calculated in Fig. 6.22.

This measurement procedure is of course cumbersome and locating leaks is difficult. Therefore, cable test equipment manufacturers have made available several types of leakage locating and detection devices. Cable operators now can, with the appropriate monitoring, locate and repair leaks, and once a year the leaks can be located, measured, and the CLI calculated. The CLI formula is shown in Fig. 6.23. The maximum allowable CLI is 64, and systems close to this number are too close and could drift in and out of tolerance, raising the possibility of severe fines. Most plant leakage detectors place a test carrier on the system at normal operating level modulated by a distinctive tone identifiable using a narrow-band sensitive receiver and a test monopole vehicular-mounted antenna. Hand-held units can be of a simple scanner type that monitors several cable-only channels or simply a separate receiver tuned to the test carrier. One instrument uses a four-level transmitter with each level modulated with a different au-

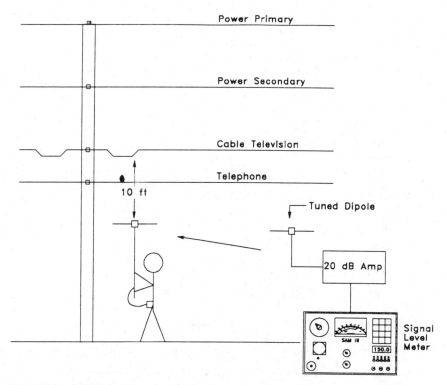

Figure 6.21 Cable television plant leakage measurement.

Numerical gain of a Dipole G = 1.64, 2.14 dB.
Distance from plant D = 10 ft, 3 m
Test Frequency 108.625 MHz
          20 microvolts/meter allowable
Allowable Maximum SLM Reading (dBmv) = $20 \log \left[\dfrac{60 \sqrt{G}}{D \times 0.026893 \times f(MHz)} \right] - 60$

$$= 20 \log (8.768) - 60$$

$$= 18.86 - 60$$

$$= -41.14 \text{ dBmV}$$ Most SLMs will not read
                               this low, therefore the
SLM reading (with amplifier) is −21.14 dBmV. 20 dB amplifier is needed.
Any reading greater is in violation.

Figure 6.22 Calculation of maximum allowable SLM reading.

Cumulative Leakage Index (CLI) = $10 \log \left[\dfrac{\text{Total Plant Miles}}{\text{Test Plant Miles}} \times \text{Sum of all leaks}^2 \right]$

CLI must be less than 64

Figure 6.23 Formula for CLI.

Normal Headend

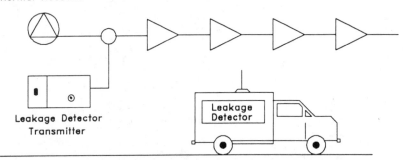

Leakage Detector
Transmitter

Leakage
Detector

Figure 6.24   Plant leakage location method.

dio tone. The highest level is at normal carrier level and the others decreasing in 5- or 6-dB steps. The worst leak will produce all four tones and a small leak only the high-level tone. This instrument makes it easier to locate the worst leaks, which are required to be repaired immediately. An example of a leakage-detection system is shown in Fig. 6.24.

Plant leakage control is something cable television operators are going to have to live with a long time. Better instruments will be developed to aid in this work. LAN operators may not have as difficult a time because many LANs are smaller, operate at levels below those used by cable systems, and often can avoid carriers in the aircraft bands.

The rules for cable systems are relaxed if the operating level is kept below the +38.5 dBmV value. However, it must be remembered that rules or not, leaking plant will cause poor performance and television service. Several businesses offer flyover services that are acceptable to the FCC for CLI certification. Special antennas and detecting equipment along with an accurate navigational system are required to identify problem areas and make the measurement. It should be remembered, a flyover was how the problem was found in the first place. However, the leaks occur on the ground along the plant routes.

6.4   Fiber-Optic Cable Communications

Cable operators and LAN contractors have discovered that a fiber-optic cable system can provide superior performance, does not leak, and is cost-effective. The optical transmitter is a laser-operated device where the light output is modulated. Actually, fiber-optic systems do not operate in the visible spectrum. However, damage to the eye can result if one looks directly into the laser output. Therefore, proper caution should be taken and protective eyeglasses worn. The receiver sys-

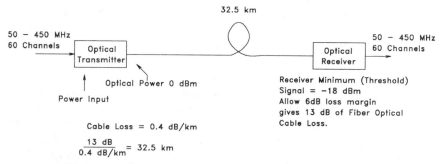

Figure 6.25 Elementary fiber-optic communication system.

tem is an optical diode and the energy from the fiber optic cable is coupled to this diode. The electrical output of the diode is the modulating
signal. Fiber-optic cable can be either multimode or single mode. For
communication purposes, single-mode fiber offers the best performance figures.

6.4.1 Fiber-optic cable parameters

Performance of a fiber-optic communication system depends on the
loss figures of the cable, the transmitter power output, and the receiver
sensitivity. These parameters control the span distance of a segment
of plant. For cable television use, fiber-optic plant presently operates
as a trunk or signal transportation system connected to nodes where
RF normal coaxial cable techniques are used to distribute cable television service to subscribers. An elementary fiber-optic system is
shown in Fig. 6.25.

The allowable 13 dB of cable loss is considered as the loss budget.
Various connectors have been developed which allow the glass fiber to
be connected to the external pieces of equipment. These connectors
also allow test equipment to be connected.

6.4.1.1 Optical power meter. Optical power meters are available from
a number of manufacturers. These instruments are offered in simple
hand-held or cabinet-mounted versions. Many offer measurements of
optical power at two wavelengths, 1310, 1550, 850, and 1310 nm, for
example. The levels range from −50 to +10 dBm with accuracies of
±0.25 dB. Most hand-held units have limited ranges and accuracies
may be about ±1 dB. Prices for the lowest-cost units range from $450
to $1200. Manufacturers also have available light sources that have
LEDs and are fitted with standard connectors. These light sources can
be used in conjunction with a power meter to measure losses as well as
identifying fibers. A variation to these instruments appears as a talk

set or two-way telephone where maintenance workers can communicate with each other when tracking down system problems. A spare fiber can be used for two-way communications using the talk set. A talk set can also be used to identify various fibers in a cable.

6.4.1.2 Measurement of optic power and loss.
Tests using a power meter on a laser transmitter can verify the specified output given by the manufacturers as well as define the input or transmitted optical power level. Also, at the receiver location a measurement of the receive power can determine if the receiver has the correct value of optical power. If the signal is too high, an optical attenuator may be needed. The difference between the transmitted power and the power at the receive site determines the value of the optical path loss of the fiber. Figure 6.26 shows this technique. A good decent-quality optical power meter is the basic and most-used piece of fiber-optic test equipment and is a definite necessity.

6.4.2 Fiber-optic cable system problems

Although fiber-optic cable is more rugged and lighter than coaxial cable, it is not without its problems. When the cable system is utility-pole-mounted, traffic accidents involving the utility pole plant can cause damage and cutting of the fiber-optic cable. Plant restoration is a problem faced by maintenance and repair crews. At splice locations, extra cable is formed in loops, often on each side of the splice, and hung on the strand. If a pole is hit, the extra cable breaks loose, preventing the cable from trying to support the pole plant. Also, the extra cable allows the splice to be replaced if damaged. The most common fault is some form of break in the cable.

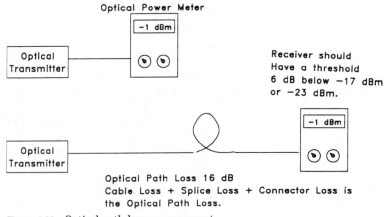

Figure 6.26 Optical path loss measurement.

6.4.2.1 The optical time domain reflectometer. An optical time domain reflectometer (OTDR) is the instrument made for finding cable faults in fiber-optic cable. Several manufacturers make this type of instrument. The Tektronix TFP2 is a fine example of this type of instrument which can measure out to 200 km with a resolution of 1 m and an accuracy of ± 0.001 percent. Mass storage of data can be had as an optional 3.5-in floppy disk drive or a high-speed printer option provides hard copy of the results. This instrument can contain two plug-in selections for the test signal wavelength. The instrument can be used for either single- or multimode fibers. A multicolor screen with popup menus with automatic help screens make up the main display. The unit is normally powered from the commercial power source with 10 to 32 Vdc power as an option. For simple multimode testing, the TEK OF150 operating at 850 nm can find faults in cables out to nearly 20 km with an accuracy of ± 0.3 percent and a resolution of 1 m. Multimode fiber systems find use in signaling and control lines and lower-speed data over shorter distances. Generally the OTDR-type instruments are expensive and cost in the range of $15,000 to $30,000. Figure 6.27 shows the fault-finding capability of the OTDR.

Another instrument that is less costly, simpler to use, and battery- or vehicular-powered is the TEK TFS2020 Optical Fault Finder, recently developed by Tektronix. The LCD screen shows a simple distance trace and an alphanumerical readout of the results. This instrument has the ability to find catastrophic cable problems, not losses and smaller reflections. Its four-button setup, two outputs, one for

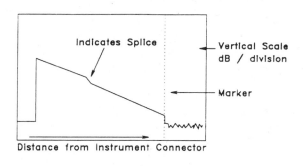

ODTR Screen

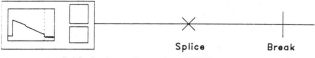

Figure 6.27   Cable fault testing using OTDR.

short and one for long range, and its 5-pound weight make this instrument attractive to technical and repair personnel.

6.4.2.2 Fiber-optic cable connectors and splicing.

Fiber-optic cable splicing and connecting is still a major problem today. Mechanical splices have more loss and are subject to coming loose and falling apart. Fusion splicing basically is a welding or fusing of the glass fibers together and is often the method of choice. Fusion splice losses are generally about half of the mechanical splice losses. However, a poor fusion may have the same loss as a good mechanical splice. Often a good fiber-optic jumper cable is used to adapt one type of connector to the other. When selecting test equipment, attention should be paid to the connector options available and should match those used in the system where the instrument will be used.

6.4.2.3 Fiber-optic cable plants.

Fiber-optic cable systems are simply a conduit for a communication medium. Some uses today are cable television service, telephone services, digital communication networks, and control system communications. The acid test of any communication system is a basic signal test from input to output. For television signals the video quality tests at the system input and output will tell the story. For computer networks, a test of the bit-stream errors or bit error rate (BER) will be a viable test. Telephone systems constantly monitor noise in their signal paths and the digital portions of the telephone networks employ BER testing. Fiber-optic cable systems are being improved daily and now instruments will soon appear with more features and at less cost.

7

Digital Systems

7.1 Digital Systems Development

Like many areas in the field of electronics, digital systems had humble beginnings. The pulse circuitry derived from radar development during World War II found new use in logic and control methods. Binary numbering and boolean algebra provided the tools needed to start the development of digital control and computational systems. Many people working in the digital and computer area of electronics today recall the diode and resistor AND and OR gates. Inverting amplifier circuits using operational-amplifier (op-amp) techniques provided signal inversion, allowing the binary 1 to be converted to a 0. Rudimentary logic circuits came into use as so-called smart controls. Further work in combining these circuits caused the NAND NOR logic to come into use. Transistor circuits as switches made possible logic circuits that operated at higher speeds. Low power consumption, high-speed operation, and relatively moderate cost made this type of logic, known as transistor- logic (TTL), attractive and still in use today. The highest-speed logic systems in use today are known as emitter-coupled logic (ECL). Complementary metal oxide semiconductor (CMOS) logic uses very little power but is slower than either TTL or ECL. Another circuit developed from the old radar days was the multivibrator, the predecessor of the flip-flop. A latch circuit made from a reset-set (R-S) flip-flop plus a clocked R-S flip-flop provided two valuable logic control circuits. Subsequently, the J-K flip-flop was developed along with a variety of other logic circuits, enabling counters of various types counting in binary or decimal (decade) to be developed. The development of shift registers using J-K flip-flops and other circuits made possible the shifting of digital pulse trains (serial) to parallel and vice-versa. With the formation of binary numbers into words or bytes, logic circuitry could be made to perform computations. Thus the digital computer evolved. Starting almost as a simple electronic desk calculator, the computer industry advanced rapidly, causing solid-state device development to advance quickly as well. Development of the integrated circuit allowed various logic circuits to contain many logic building blocks. The printed circuit board wiring connecting the logic building blocks made possible higher operating speeds. Integrated circuit logic devices are available in simple form or in large-scale integrated (LSI) circuits that are operated as a computer subsystem known as a microprossessor. Further solid-state development produced very large scale integrated (VLSI) circuits, again speeding up the logic functions needed to develop faster computers. The first part of this chapter is a study of the digital quantities and devices that will be measured and tested.

7.1.1 The binary numbering system

The series of pulses of voltage or current can represent the ON (binary 1) or OFF condition (binary 0). If the ON condition is a positive voltage or current and the OFF condition a zero voltage or current, then this is known as *positive true logic*. If the reverse is true, then it is called *negative true logic. The presence of a binary 1 is often called a true* and the zero (off) condition a *false.* A table of the output condition of a circuit is known as a *truth table.* This table enables one to trace an output terminal's response to the input stimulus. A plethora of logic symbols came into use so that digital designers could document the designs and enable testing and troubleshooting procedures to be developed. A few of the common logic circuit symbols with their truth tables are shown in Figs. 7.1 through 7.3. These logic functions are combined into systems according to a logic diagram, which in turn is used in the design of the printed circuit boards. The interconnections between the various logic functions are used to make systems that are able to perform various arithmetic and control operations. Several functional logic circuits appear as integrated circuits that perform such operations as decimal decoders and display. According to binary arithmetic, it takes three binary bits to represent the eight decimal digits 0 to 7. In order to represent ten decimal digits, four binary bits are needed. A simple binary-to-decimal decoder is shown in Fig. 7.4.

A two-state logic device called a flip-flop is also in wide use today. Several types are available which act as latching or gate devices. Three more popular flip-flops which will be studied are the simple reset-set (R-S) flip-flop, the D latch flip-flop, and the J-K flip-flop. The logic function symbols and the truth tables for these three devices are shown in Fig. 7.5. Flip-flop logic devices are extensively used in

Symbol   Truth Table

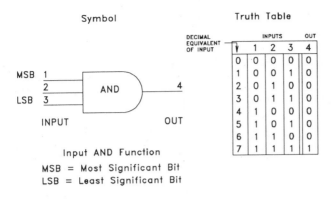

DECIMAL EQUIVALENT OF INPUT	INPUTS			OUT
	1	2	3	4
0	0	0	0	0
1	0	0	1	0
2	0	1	0	0
3	0	1	1	0
4	1	0	0	0
5	1	0	1	0
6	1	1	0	0
7	1	1	1	1

Input AND Function

MSB = Most Significant Bit
LSB = Least Significant Bit

All Inputs must be True (Binary 1) in order
to have a True Output.

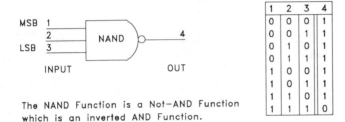

1	2	3	4
0	0	0	1
0	0	1	1
0	1	0	1
0	1	1	1
1	0	0	1
1	0	1	1
1	1	0	1
1	1	1	0

The NAND Function is a Not–AND Function
which is an inverted AND Function.

Figure 7.1 AND, NAND function.

counters, registers, storage elements, and clocking devices in digital circuitry.

Logic circuits respond to single voltage levels (a voltage step) or a pulse train consisting of logical 1s and 0s. The coding and decoding of binary numbers is often a useful tool in diagnosing problems in digital circuitry. Table 7.1 shows the binary equivalent to several numbering systems. It should be recalled that the ASCII code was mentioned in Chap. 6, Fig. 6.3 and Table 6.2.

The pulse trains of digital signals are sent through communication lines using several carrier modulating schemes. Frequency shift keying (FSK) of a carrier and biphase modulation are often used. Groups of binary numbers are often arranged in a pulse-code modulation (PCM) scheme and then transmitted through a modulated carrier. A simple explanation of an example of PCM is the rotation of a wheel, where the start of a rotation is marked by a specially designated group of binary bits. Following this group would be, say, 16 bits of data belonging to message 1 with the next 16 bits belonging to message 2,

Symbol   Truth Table

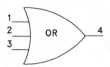

1	2	3	4
0	0	0	0
0	0	1	1
0	1	0	1
0	1	1	1
1	0	0	1
1	0	1	1
1	1	0	1
1	1	1	1

The OR Function--any True Input will
cause a True Output

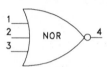

1	2	3	4
0	0	0	1
0	0	1	0
0	1	0	0
0	1	1	0
1	0	0	0
1	0	1	0
1	1	0	0
1	1	1	0

The NOR Function is a Not-OR
Function, which is an Inverted
OR Function

Figure 7.2   or, nor function.

Symbol   Truth Table

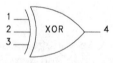

1	2	3	4
0	0	0	0
0	0	1	1
0	1	0	1
0	1	1	0
1	0	0	1
1	0	1	0
1	1	0	0
1	1	1	0

The Exclusive Or function requires only one
input to be true for a true output.

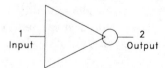

1	2
0	1
1	0

The Inverter function changes a true (1)
input to a false (0) output and vice versa.

Figure 7.3   Exclusive or and the inverter function.

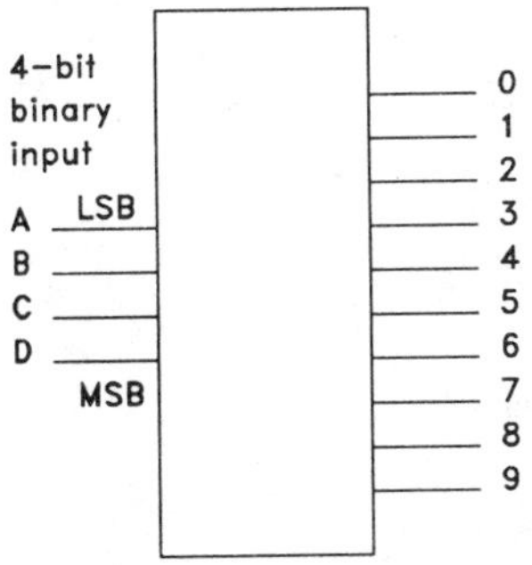

	Input			Output									
MSB			LSB										
D	C	B	A	0	1	2	3	4	5	6	7	8	9
0	0	0	0	0	1	1	1	1	1	1	1	1	1
0	0	0	1	1	0	1	1	1	1	1	1	1	1
0	0	1	0	1	1	0	1	1	1	1	1	1	1
0	0	1	1	1	1	1	0	1	1	1	1	1	1
0	1	0	0	1	1	1	1	0	1	1	1	1	1
0	1	0	1	1	1	1	1	1	0	1	1	1	1
0	1	1	0	1	1	1	1	1	1	0	1	1	1
0	1	1	1	1	1	1	1	1	1	1	0	1	1
1	0	0	0	1	1	1	1	1	1	1	1	0	1
1	0	0	1	1	1	1	1	1	1	1	1	1	0

A zero appearing in the otput column indicates the active decimal digit.
Note: Binary numbers beyond 1001 will indicate all output 1's.

Figure 7.4   Binary-to-decimal decoder.

and so on. After, for example, 10 message words, a group of binary bits containing the total number of bits sent in the 10-word frame can be used for accuracy checking. Following this so-called parity word would be the next frame start word to repeat the process. At the receiving end the demodulated signal will drive the decoding circuits, which find the frame-synchronizing point and separate the various message words to assemble complete messages (1 to 10) for routing to the receiving workstations. The telephone industry converts telephone analog signals to equivalent digital numbers, assembles the PCM pulse train signal containing many telephone conversations for transmission through the telephone lines to terminals which demodulate and decode each conversation and route the telephone message to the receiving party.

7.1.2 The digital word or byte

The digital world contains so many buzzwords and abbreviations that sometimes it is difficult to understand what is going on. Often, certain terms are used synonymously with others, thus causing confusion. The terms "data word" or "digital word" and the term "byte" constitute one such example. The main difference between a word and a byte is that a word is a complete unit of measure, that is, a number or other symbol. A computer word consists of a number of bits, for instance, an 8- or 16-bit word. A *byte* usually is defined as a number of bits. Therefore, an 8-bit byte could consist of an 8-bit word or two bytes could consist of one 16-bit word. Data-handling systems like computers and microprocessors usually work in bytes, the most common being the 8-bit byte, which is able to represent the characters A to Z. The ASCII character consists of an 8-bit byte.

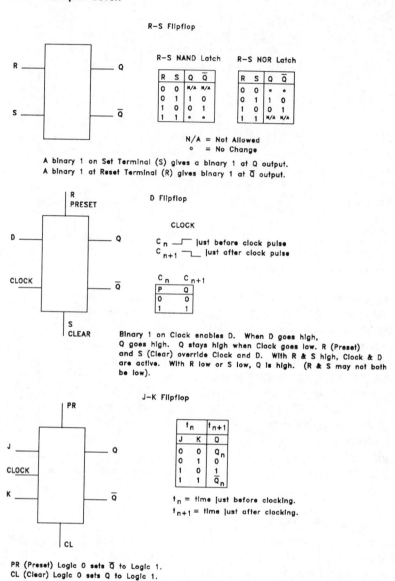

R–S Flipflop

R–S NAND Latch

R	S	Q	Q̄
0	0	N/A	N/A
0	1	1	0
1	0	0	1
1	1	°	°

R–S NOR Latch

R	S	Q	Q̄
0	0	°	°
0	1	1	0
1	0	0	1
1	1	N/A	N/A

N/A = Not Allowed
° = No Change

A binary 1 on Set Terminal (S) gives a binary 1 at Q output.
A binary 1 at Reset Terminal (R) gives binary 1 at Q̄ output.

D Flipflop

CLOCK

C_n ⎍ just before clock pulse
C_{n+1} ⎎ just after clock pulse

C_n	C_{n+1}
P	Q
0	0
1	1

Binary 1 on Clock enables D. When D goes high,
Q goes high. Q stays high when Clock goes low. R (Preset)
and S (Clear) override Clock and D. With R & S high, Clock & D
are active. With R low or S low, Q is high. (R & S may not both
be low).

J–K Flipflop

t_n		t_{n+1}
J	K	Q
0	0	Q_n
0	1	0
1	0	1
1	1	$\bar{Q}_n$

t_n = time just before clocking.
t_{n+1} = time just after clocking.

PR (Preset) Logic 0 sets Q̄ to Logic 1.
CL (Clear) Logic 0 sets Q to Logic 1.

Figure 7.5 Some common flip-flops with truth tables.

7.1.3 Development of digital instruments and controls

Many pieces of electronic equipment, although analog in principle, have digitally operated controls. Push-button controls are often operated in the digital domain while controlling analog devices. Many industrial manufacturing processes used mechanical counters to log the

TABLE 7.1   Some Common Digital Codes

Binary	Decimal	BCD	Octal	Hexa-decimal	ASCII
00000	00	0000 0000	000	00	00110000
00001	01	0000 0001	001	01	00110001
00010	02	0000 0010	002	02	00110010
00011	03	0000 0011	003	03	00110011
00100	04	0000 0100	004	04	00110100
00101	05	0000 0101	005	05	00110101
00110	06	0000 0110	006	06	00110110
00111	07	0000 0111	007	07	00110111
01000	08	0000 1000	010	08	00111000
01001	09	0000 1001	011	09	00111001
01010	10	0001 0000	012	0A	Need 2 ASCII
01011	11	0001 0001	013	0B	Nos.
01100	12	0001 0010	014	0C	
01101	13	0001 0011	015	0D	
01110	14	0001 0100	016	0E	
01111	15	0001 0101	017	0F	
10000	16	0001 0110	020	10	
10001	17	0001 0111	021	11	
10010	18	0001 1000	022	12	
10011	19	0001 1001	023	13	
10100	20	0002 0000	024	14	
10101	21	0002 0001	025	15	
10110	22	0002 0010	026	16	
10111	23	0002 0011	027	17	
11000	24	0002 0100	030	18	
11001	25	0002 0101	031	19	
11010	26	0002 0110	032	1A	
11011	27	0002 0111	033	1B	

amount of units completed, and since mechanical things seemed to wear out, electronic digital equipment was developed. A photocell and light source replaced the mechanical counting device. The pulses from the photocell connected to a counting circuit made up of flip-flops and associated logic gates. Integrated circuits accumulated the binary count and converted to decimal numbers representing the total per the desired time span. Accumulating the total every minute will give the rate of units or operations per minute. These results could be displayed on an operating panel, stored in a solid-state random access memory (RAM) for future computer analysis, or used for real-time corrections to the manufacturing process. Many manufacturing processes operate in this fashion, particularly the oil and gas industry, the food processing industry, the brewing and distilling industry, and the pharmaceutical industry. Essentially the recipe is stored in a digital memory while the analog sensor outputs are converted to digital numbers by devices known as analog-to-digital converters (ADCs). These sensors have a voltage output proportional to the temperatures, pressures, and/or flows necessary to monitor the process. These measurements are compared against the correct values in the memory cells and the errors are reported on a control panel display. Most modern-day process control systems use a feedback technique which uses the error measurements to control the valves, and system controls to correct the overall process to conform to the stored recipe. The digital error signals are converted using digital-to-analog (DAC) circuits, which cause motors and actuators to correct the process.

7.1.3.1 A/D and D/A conversion. An important part of most digital controls systems is the ADC and the inverse DAC. The ADC operates on the principle of sampling the analog signal at a rate at least five times the highest-frequency component present in the input signal. These samples are held sufficiently to allow time to generate a digital number proportional to the amplitude of the sample. These numbers are then routed to circuits that store this digital number for more operations. The inverse of this operation is performed by the DAC, where digital numbers are summed from the most significant bit (MSB) to the least-significant bit (LSB). The resulting electrical current is proportional to the value of the digital number. The DAC output should be able to reproduce an accurate replica of the input signal. The number of binary bits per sample and the resolution of the LSB determine how close the DAC output is to the original analog signal. A diagram of an elementary DAC is shown in Fig. 7.6. This system shows an 8-bit word of digital data, which allows the analog signal to be divided up into 256 total bits. A 1-V signal divided into 256 samples will give a resolution of $\frac{1}{256}$, which is 0.0039 V or 3.9 mV for the LSB. Some ADCs use a DAC as part of the circuit. The successive-

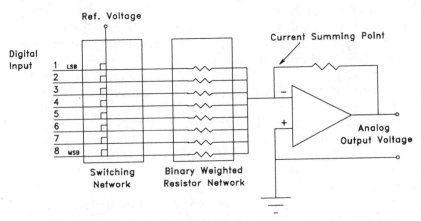

Figure 7.6 Elementary digital-to-analog converter.

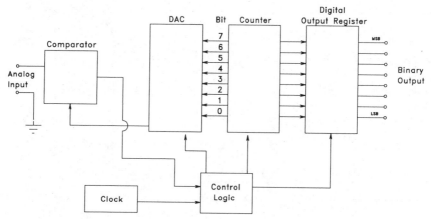

Figure 7.7   Elementary analog-to-digital converter.

approximation ADC uses the output of a DAC which is driven from an internal counter and clock to compare the output voltage to the analog input voltage, starting with the MSB and progressing to the LSB. The binary bit is retained if the analog input voltage is less than the DAC voltage. This basic ADC is shown in Fig. 7.7. The ADCs used in many digital voltmeters appear in a single-chip configuration complete with counter circuits, control logic, and display circuitry. The use of ADCs and DACs allows digital techniques to live in an analog world. Many control functions in manufacturing are analog in nature. Converting these parameters to digital quantities allows the data to be stored in digital memories for future analysis or to generate control signals used to correct the process. Digital data are often used in alphanumerical displays such as the LED, the LCD, and fluorescent digital displays commonly used in appliances, radios, television sets, and automotive equipment. A diagram for one number using the common seven-segment LED is shown in Fig. 7.8. Other digital display devices operate in a similar fashion with control lines used to activate the various areas of the device. The decoder circuits appear in a variety of single integrated circuit (IC) packages available from many manufacturers. The IC field in general is so large that catalog information could fill a room. A set of computer disks is available with most IC operating parameters and specifications needed by today's circuit designers.

7.1.3.2 Microprocessors.   Many of today's control systems use either a microprocessor, microcomputer, or minicomputer as the main control element. Even the personal computer (PC) is used as an instrument or control device, especially when speed and analysis are important. Microprocessors and what are known as single-chip computers have to

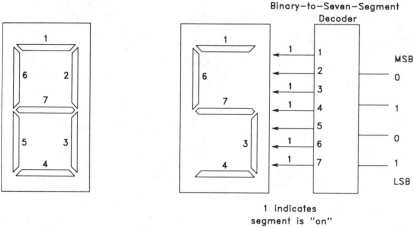

Figure 7.8 Seven-segment display.

operate under some programming device. Sources of programming often appear as a single-chip programmable memory (PROM) or the erasable programmable memory (EPROM). The program information stored in these memory devices instructs the microprocessor on what it is supposed to do and how to react to input/output digital data. An external device known as a *PROM burner* is used to program the device, and once programmed, it cannot be changed. If changes are necessary another PROM chip has to be prepared. The EPROM, on the other hand, can be erased by an ultraviolet source of energy and reprogrammed using a PC-controlled programming station. Exposing the open window on the device to sunlight can be rudimentary form of erasing. However, a device is available that provides the proper amount of ultraviolet light necessary for complete erasure. An adhesive covering is applied over the window before the device can be reprogrammed. Once the device is programmed, a label can be attached to identify the internal program.

Microprocessor control systems can take on many different roles. Some appear in manufacturing process controllers, environmental controllers, and engine system controllers, to name a few. Aircraft, marine, and automotive engine systems often employ microprocessor controllers. A simple system for temperature control is shown in Fig. 7.9. Microprocessor-microcomputer systems vary in size and complexity. Before trying to test one of these devices, it is very important to obtain the performance data from the manufacturer's published data. Often manufacturers publish application notes which contain a wealth of information as to uses and applications, along with any warnings or parameters not to be exceeded.

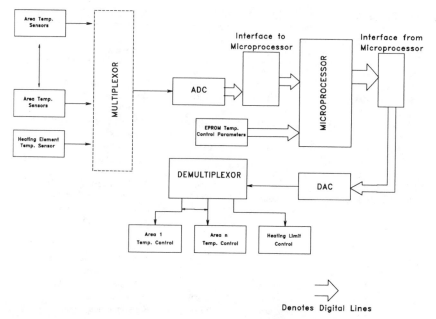

Denotes Digital Lines

Figure 7.9   Example of a microprocessor temperature control system.

7.1.3.3 Computer and microprocessor equipment and standards. Computers and microprocessor equipment usually have to be connected to the outside world through peripheral equipment. Since microprocessors run on a clocking system, the clock function may have to be passed through to other equipment if synchronous operation is necessary. The communication lines running between microprocessor controllers and the peripheral equipment are called bus lines or simply buses. Probably the most common bus is the RS-232, which has been adopted as an international standard. Digital data are carried in 8-bit words in a serial fashion. The 8 bits conform to the ASCII code as shown in Fig. 6.3 and Table 6.2. The standard 25-pin D-type connector is used, and the pin assignments are given in Table 7.2.

A one-to-one flat ribbon cable with a male connector on one end and a female connector on the opposite end is used to interconnect the modem to the terminal. The EIA RS-449 standard using a 37-pin connector is being accepted more and more in many applications. Schematic diagrams and wiring prints of systems used today will give pin assignments and test points. The RS-232C standard has available the necessary control and data signals to use a dial-up modem or a dedicated on-line modem for transmission and reception of data from and to the terminal. Modems in use today allow data to be transmitted and received over the telephone system. Using the worldwide telephone system, computers around the globe can communicate with each other.

TABLE 7.2   Terminal-to-modem pin connections for RS-232C 25-pin D-type connector

Pin no.	Circuit mnemomic	Source	Type	Remarks
1	AA	Ground	Ground	Chassis.
2	BA	Terminal	Data	Terminal sends data.
3	BB	Modem	Data	Modem sends data to terminal.
4	CA	Terminal	Control	Request to send from terminal.
5	CB	Modem	Control	Modem clean to receive from terminal.
6	CC	Modem	Control	Data set is ready.
7	AB	Ground	Ground	Signal ground.
8	CF	Modem	Control	Carrier detect.
9	—	—	—	Positive test lines.
10	—	—	—	Negative test lines.
11	—	—	—	Not used.
12	SCF	Modem	Control	Secondary carrier detect.
13	SCB	Modem	Control	Secondary clear to send.
14	SBA	Terminal	Data	Secondary data transmit.
15	DB	Modem	Timing	Transmit clock + edge.
16	SBB	Modem	Data	Secondary receive data.
17	DD	Modem	Timing	Receive clock edge.
18	—	—	—	Not used.
19	SCA	Terminal	Control	Secondary request to send.
20	CD	Terminal	Control	Data terminal ready.
21	CG	Modem	Control	Signal quality off (error).
22	CE	Modem	Control	Ring indicator.
23	CH/CI	Terminal or modem	Control	Data signal rate selector.
24	DA	Terminal	Timing	Transmit clock edge.
25	—	—	—	Not used.

Higher-speed modems are available to send and receive data through dedicated microwave and satellite communication systems. For short distances a modem is not necessary. Such applications as simply connecting a printer or other peripheral piece of equipment in the same room just requires the standard RS-232C cable set. For such an application, not all pin assignments are used and a short form of cabling results, as shown in Fig. 7.10. Many PCs have a separate parallel printer port making such cabling unnecessary. Most equipment using the RS-232C standard interconnection use an IC to change the digital serial data stream to parallel for efficient computer operations. The ICs are usually 40-pin universal asynchronous receiver transmitters (UARTs), which receive and or transmit data in the usual unsynchronized 8-bit words or bytes. The necessary handshake for each 8-bit character of data causes this transfer method to be relatively slow. Universal synchronous/asynchronous receiver transmitters (USARTs) allow significantly faster synchronous operation. Modems

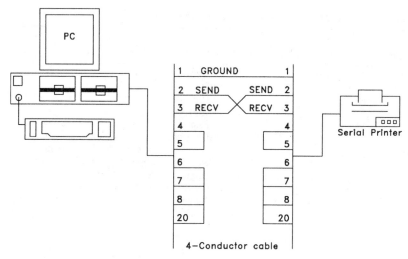

Figure 7.10   Short-distance RS-232C cabling.

act as the RS-232C interface to the telephone system and operate by modulating a carrier, for example, by FSK or some other method, such as biphase keying. Use of modems allows digital communications between terminals located in distant parts of the world.

Another bus in wide use today is the general-purpose interface bus (GPIB), which conforms to IEEE 488 standard. This bus architecture allows equipment to be connected for a distance of about 20 m. Longer distances between equipment are possible by using a bus extender. The bus extender essentially terminates the cable and amplifies the signal by acting as a repeater. Distances of 100 m between equipment can be reached using the bus extender. The GPIB interface uses a 24-pin connector, where data are transmitted in the parallel mode in 8-bit bytes in the non- or asynchronous mode. Even though the connector is 24 pins, only 16 wires are needed to carry the data, the handshakes, and the bus management lines. The 16 lines and connector pin assignments making up the IEEE 488 bus are given in Table 7.3.

The IEEE 488 GPIB bus is mainly used in instrument control and data logging of the measured values. Many automatic testing systems contain instruments using the GPIB interface coupled to a controlling computer. Tektronix Corp. has a variety of instruments with the ability to be GPIB-controlled, along with the necessary software to control the whole testing program. The Tektronix special codes and formats which conform to the proposed IEEE P981 standard avoid some of the system processor confusion. It is advisable when considering using a computer-controlled testing system for a particular application to solicit advice from the instrument manufacturer's application or customer engineering department. Tektronix has available the EZ-TEK

TABLE 7.3   IEEE 488 GPIB Connector Pin Assignments

Pin assignment	Signal	Remarks	Type
1	Data 1	Digital data	
2	Data 2	Digital data	
3	Data 3	Digital data	
4	Data 4	Digital data	
5	EOI	End of identify	Bus management
6	DAV	Data valid	Handshake
7	NRFD	Not ready for data	Handshake
8	NDAC	Not data accepted	Handshake
9	IFC	Interface clear	Bus management
10	SRQ	Service request	Bus management
11	ATN	Attention	Bus management
12		shield	
13	Data 5	Digital data	
14	Data 6		
15	Data 7		
16	Data 8		
17	REN	Remote enable	Bus management
18–23		Ground	
24		Logic ground	

software package which can be used either with the TEK PEP 301 controller or a compatible IBM-type PC fitted with an IEEE 488 standard GPIB card. An instrument such as the TEK 2465A oscilloscope with option 10 (GPIB) interface can be used to perform automatic measurements and data logging under computer control.

There are many more bus interconnection systems in use today. However, they all more or less operate in the same fashion and some are used more widely than others. The RS-232C and the IEEE systems have been in use for many years and are still in wide use today. Hence, they both make good examples of bus interconnection methods between instruments, computers, and control systems.

7.2   Digital Test Equipment

Test equipment for digital systems range from the very simple to the extremely complex. Such types of instruments are simple logic probes and logic pulsers to oscilloscopes, counters, logic analyzers, and system analyzers. Insertion of break-out boxes in bus lines facilitate use of either logic probe oscilloscopes or logic analyzers to analyze the signals on the bus lines. The level of testing ranges from simple logic testing of integrated circuits to complete system testing. Computer and microcomputer systems can be analyzed by diagnostic programs that may indicate in which area the problem is located. Substitution of a faulty sub-system will return the computer system to proper on-line operation. Board testers under microcomputer control with diagnostic software can often pinpoint the faulty integrated circuit.

7.2.1 The logic probe, logic pulser

Two of the simplest and lowest-cost instruments used in the digital domain are the hand-held logic probe and the logic pulser. The logic probe is usually powered off the circuit being tested using polarity-marked clip leads. The logic probe will give an indication by lighting LEDs as to the logic state, that is, a logic high or logic low corresponding to a binary 1 or 0. Also, a switch and LED on the probe will indicate the presence of a pulse-train condition such as a clocking function. Most logic probes have a selector switch for the two most popular logic circuits, TTL or CMOS. Some logic probes have an audio tone that indicates the high or low condition. Usually, the higher-pitched tone indicates a high logic condition and vice versa. Use of a logic probe is demonstrated in Fig. 7.11.

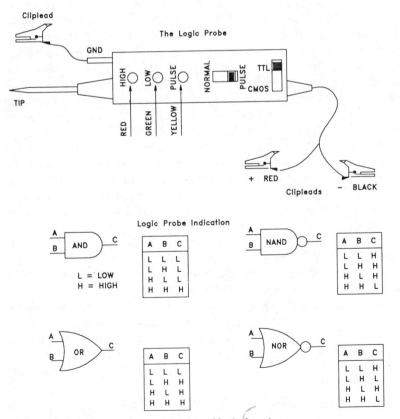

Figure 7.11 The logic probe and testing of logic functions.

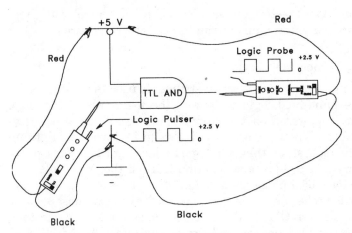

Figure 7.12 Logic pulser, logic probe test of AND function.

The hand-held logic pulser looks much like the logic tester and is powered from the circuit or uses internal batteries. A selector switch chooses pulses for TTL or CMOS logic levels. This instrument is handy for testing circuits mounted on a breadboard in a system where inputs to the logic functions are known. If a logic pulser and probe are properly powered, touching the probe tips together will activate the pulse indicator light on the logic probe when the switch is in the pulse condition. This procedure can be useful in testing printed circuit board continuity. Figure 7.12 shows use of the logic probe and pulser in the testing of an AND logic circuit. This type of testing is conducted on the printed circuit card containing the integrated logic circuits or through the means of a breakout box which allows testing to take place on an interconnecting cable. Breakout boxes are usually available as a flat ribbon or other cable types with matching connectors on each end. In the middle the cable conductors are exposed for logic testing. Breakout boxes are available with jumper links allowing a conductor in the cable to be broken so that both ends can be independently tested. Some breakout boxes are not really breakout boxes in that the cable is not exposed for testing and a series of LEDs is connected to the conductors to indicate pulsing and/or logic state. The most common breakout boxes are for RS-232C type.

Similar to a breakout box is a loop-back tester, which in essence connects the bus output for a device back to the same device's input. In the middle, LEDs indicate the status of the control lines. Most loop-back testers are special types for specific bus methods.

7.2.2 The logic analyzer

Early in the development of digital circuitry, the oscilloscope was a most-valued piece of equipment. The structure of electrical pulses could be analyzed as to rise time, fall time, pulse duration, and amplitude. Multi-input and -output logic circuits had to be analyzed piecemeal. Consider the fact that a simple two-input AND gate will require three oscilloscope traces to simultaneously look at both inputs and the output. Multiple-trace oscilloscopes with limited amplitude per trace were developed and were the forerunners of the present-day logic analyzer. Some of these logic analyzers are still in use and are particularly good for examining the system timing pulses. However, looking at scope traces on eight digital lines can be a problem, as can remembering which line belongs to which trace. The Tektronix model 1230 logic analyzer is an extremely useful instrument for logic analysis with four multipin logic probes and built-in software. The keypad on the front panel allows the functions to be selected from screen pop-up menus. The traces are identified by screen labeling at the input lines and showing the logic functions and timing on 16 traces at once. Data can be stored in four nonvolatile memory banks. Data from two memories can be compared, which could indicate a fault between measured input and output data. A variety of cable probes allows logic testing between devices. An RS-232C or IEEE 488 GPIB allows programming, data logging, data retrieval, and analysis on standard IBM PC equipment. Options include RS-232C and GPIB communication interfaces. These cards can be user-installed at a later date. A 100-MHz digitizing oscilloscope card is an extremely useful option. Since this instrument is very modular with many options, its use and application should be discussed with Tektronix's applications engineering staff. A tailor-made instrument can be put together that will perform the required logic analysis at the least possible cost.

The top-of-line Tektronix Prism 3000 system offers even more options and versatility than the TEK 1230 series. This type of system is useful in checking out computer main circuit boards and complicated computer control systems. A color monitor, keyboard controls, and extensive built-in software allow testing to 2-GHz for timing functions plus many other computer and logic operations. This instrument can be many tools in one and its modular construction allows it to be a custom-built instrument as well.

Since digital testing consists of identifying the binary-coded signal levels of zeros and ones, it may seem quite simple. However, the loss of one bit can essentially cause a catastrophe, particularly if it occurs in random fashion. Data lines are often analyzed as groups of words or numbers translated by the instrument and displayed on a screen. Logic analyzers available today can deal with a large amount of digital data performing a simple task many times and at high speed. Present-day logic analyzers have alphanumeric screen displays as well as traces showing timing information. Often the screens are in multicolors to clarify the data information.

7.2.3   Digital communications testing using BERT and data analyzers

Digital communications is used extensively today transporting a multitude of digitally encoded telephone conversations as well as digital data for financial and commercial institutions. Government use of the commercial digital communication systems often has the messages encrypted (scrambled) using many digital algorithms. Testing of these communications links can be done on a partial or complete end-to-end basis. The method of choice today seems to be to test the bit error rate (BER), since the loss of even one bit can cause a sizable error. The BER tester (BERT) is used to check out the end-to-end results of a known digital data stream. Specialized BERT instruments are made for specific types of digital communication systems. For instance, there are BERT instruments manufactured for telephone industry T-1 standard digital communication systems. Also, the telephone industry has developed the digital standards called DS0, DS1, DS2, and DS3; BERT instruments are available that can analyze all four digital standards. These instruments can be placed at the end of a line or inserted anywhere along the line. Data analyzers and protocol analyzers are other specialized pieces of digital communications test instruments. The data analyzer is able to examine whole or parts of messages for correct data signals. The protocol analyzer is very similar and usually tests for proper data synchronization and BERs as well. Use of the data or protocol analyzer for T-1 link testing is shown in Fig. 7.13.

A qualitative type of digital system testing can be done with a high-frequency oscilloscope or some of the new digital oscilloscopes on the market today. In some cases where digital data are placed on a wire, line distortion of the bit stream will result. Still, the bit stream can be reconstructed accurately, provided the distortion is not too great. Also,

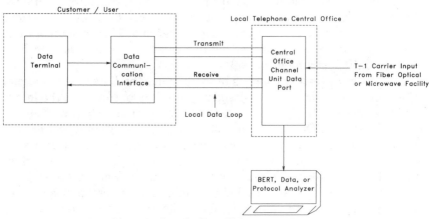

Figure 7.13 Analyzer placement for telephone T-1 system.

in some cases the digital data can be predistorted and hence will pass through the data lines without further distortion. Examination of a data stream with an oscilloscope will produce a pattern on the screen that resembles the human eye and hence is called an *eye pattern*. When the eye pattern looks like a wide-open eye, the distortion is minimal. If the eye pattern closes, distortion is great enough, so some bits will not be recognized and data errors result. If the errors are accumulated over a time period, then the BER can be calculated. Digital storage and color display tubes on some oscilloscopes enhance analysis of the eye patterns. An example of the development of the eye pattern is shown in Fig. 7.14. This method gives a qualitative analysis, whereas an actual BERT instrument will be more quantitative. However, the eye pattern can be used to locate possible areas where the distortion is occurring.

Many special data analyzers and protocol analyzers are available for local-area-network (LAN) and or wide-area-network (WAN) testing and maintenance that conforms to the large variety of standards such as RS-449, Bell 303 V.25, X.25, and RS-232C/V.24. Unfortunately, there is no do-all type of instrument available, and if there

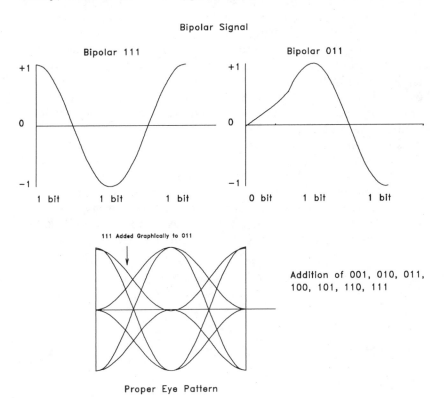

Figure 7.14 Formation of an eye pattern.

were, it would probably not do everything quickly and well. Since LAN/WAN systems may operate from simple twisted-pair cable, baseband coaxial cable, modulated FDM coaxial cable, and fiber-optic cable, the proper cable plant test equipment should be available as well as overall link test gear such as BERT instruments, data analyzers, and protocol analyzers. Marginal operation may be spotted by the link analyzers; but to find the cause, other system tests may be required.

7.2.4   Software-controlled diagnostic tests and advanced methods

Most instruments used on digital systems are often called "smart" in that they are similar in many respects to the computerized equipment they were intended to test. These instruments are usually programmable either by internal microprocessor control or interfaced to an external computer through an interface. The software is in essence the diagnostic control element. Most operating computer systems can be tested using diagnostic programs which can test the memory banks, such as random access memory (RAM), floppy disks, and hard-disk drives. Other parts of the computer system, such as the monitor, keyboard, and other peripheral equipment, and the interfaces can also be tested using diagnostic software. When a piece of equipment is suspected of being faulty, substitute testing with equipment known to be good will confirm whether it is faulty or not. Diagnostic software is designed to find which piece of equipment is faulty, and in the case of the hard-disk drive, which area of a disk is turning up errors. Most computer equipment needs to be cleaned periodically to remove dust, and often the control printed circuit cards that connect to "mother" boards need the edge connectors cleaned of corrosion and dirt.

Having computer equipment control test equipment via software is an extremely versatile method because the computer installed with a telephone modem can transmit data to remote location and receive program instructions used to control the testing operation as well. One such method is shown in Fig. 7.15. A computer could also be placed at the remote site containing the control software and an anal-

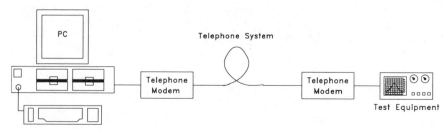

Figure 7.15 Computer-controlled test method.

ysis program as well. Processed data could then be sent to the main control computer, which in turn could supervise many other similar test setups. Whole manufacturing processes are controlled in this manner, where any measurement errors can interrupt certain operations until corrective action is taken.

When testing any digital system or data communications system, a study of the system manuals should be the first order of business. One should be very familiar with the equipment and any precautions before proceeding with any system testing. Most system operating manuals have sections on maintenance and troubleshooting procedures. Large systems often include some instrumentation, tools, and spare equipment as part of the system package. Also, extensive testing manuals accompany the delivered equipment. Most test equipment manu-

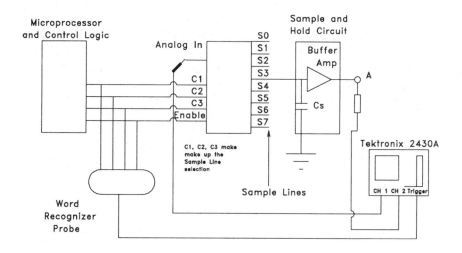

Word Recognizer Probe recognizes when Sample 3 (S3) is active

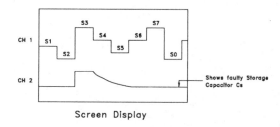

Screen Display

Oscilloscope Used with Word Recognizer Probe for Sweep Trigger

Figure 7.16 Oscilloscope used with word recognizer probe for sweep trigger.

facturers have available some form of operational and maintenance training to their customers' personnel.

When using a oscilloscope to test the operation of digital control and data systems, the selection of the oscilloscope trigger method is important. Often triggering the sweep using the system clock signal will suffice. However, in data systems where the control signal is a digital number consisting of several bits, some method of triggering the oscilloscope on the digital word sync is needed. A word recognizer probe available for the Tektronix 2430A digital oscilloscope will solve this problem. The testing of an analog multiplexer using this technique is shown in Fig. 7.16. The word recognizer probe synchronizes the oscilloscope sweep, so channel 2 trace will show the analog output for sample 3 (s3). If a word recognizer probe (option for the 2430A) is not available, a logic analyzer can provide a trigger for the oscilloscope by finding the correct control word. However, this is an example of driving tacks with a sledgehammer. For the digital design engineer, a self-constructed word recognizer probe could be built. The word selection control could be programmed by an 8-bit DIP switch. The digital oscilloscope is a universal instrument and can do many jobs in the analog and digital domain. For general applications it should be the instrument of choice.

The XYZ's of Using a Scope*

Introduction

When you watch an electrical engineer tackle a tough design project or a service engineer troubleshoot a stubborn problem, you'll see them find a scope, fit probes or cables, and start turning knobs and setting switches without ever seeming to glance at the front panel. To these experienced users, the oscilloscope is their most important tool, but their minds are focused on solving the problem, not on using the scope.

Making oscilloscope measurements is second nature to them. It can be for you too, but before you can duplicate the ease with which they use a scope, you will need to learn about the scope itself—both how it works and how to make it work for you.

This primer can help you learn enough about oscilloscopes and oscilloscope measurements that you will be able to use these tools quickly, easily and accurately. The text is divided into two parts.

Part I, in its first four sections, describes the functional parts of the scope and the controls associated with each function. It ends with a section on probes.

Part II builds on the knowledge and experience gained in Part I. The first section shows you typical signals you'll see on the oscilloscope screen and defines terms for parts of waveforms that are discussed. The next two sections cover safety and instrument set-up procedures. Then measurement techniques are described and exercises are included to help you practice a few basic measurements. You'll also find several examples of advanced techniques that can help you make more accurate and convenient measurements. The last section discusses oscilloscope performance and its effects on your measurements.

Part I. Scopes, Controls, and Probes

You can measure almost anything with the two dimensional graph drawn by an oscilloscope. In most applications the scope shows you a

---

*Used with permission by Tektronix, Inc.

graph of voltage (on the vertical axis) versus time (on the horizontal axis). This general-purpose display presents far more information than is available from other test and measurement instruments such as frequency counters and multimeters. For example, with a scope you can find out how much of a signal is direct (dc), how much is alternating (ac), how much is noise (and whether or not the noise is changing with time), what the frequency of the signal is, and more. With a scope you can see everything at once rather than having to make many separate tests.

Most electrical signals can be easily connected to the scope with either probes or cables. Transducers, which change one kind of energy into another, are available for measuring nonelectrical phenomena. Speakers and microphones are two examples of transducers. A speaker converts electrical energy to sound waves and a microphone converts sound to electricity. Other typical transducers can transform mechanical stress, pressure, light, or heat into electrical signals.

Given the proper transducer, your test and measurement capabilities with an oscilloscope are almost endless.

Making measurements is easier if you understand the basics of how a scope works. You can think of the instrument in terms of the functional blocks illustrated in Fig. A.1—vertical, trigger, horizontal, and display.

The vertical system controls the vertical axis of the graph. Any time the electron beam that draws the graph moves up or down, it does so under control of the vertical system. The horizontal system controls the left-to-right movement of the beam. The trigger system determines when the oscilloscope begins drawing by starting the horizontal sweep across the screen. And the display system contains the cathode-ray tube (crt), on which the graph is drawn.

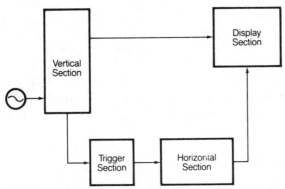

Figure A.1   The basic oscilloscope in its most general form has four functional sections (or systems): vertical, horizontal, trigger, and display. The display system is also sometimes called the *crt* (cathode-ray tube) *section.*

Part I of this appendix is divided into five sections—one for each of the four functional systems and one for probes. In each section, the controls are identified, and you can locate them on the illustration of a Tektronix 2225 Oscilloscope front panel at the back of the appendix. The controls and their functions are described, and at the end of the section, there are hands-on exercises using those controls.

The last section in this part describes probes. When you finish reading Part I, you'll be ready to make fast and accurate oscilloscope measurements.

Before you turn on your scope, remember that you should always be careful when working with electrical equipment. Observe all safety precautions in your test and measurement operations. Use the proper power cord and correct fuse. Always plug the power cord of the scope into a properly wired receptacle before connecting your probes or turning on the scope. And don't remove the covers and panels of the scope.

Now look at the photograph of the Tektronix 2225 Oscilloscope at the end of this appendix. Follow Exercise 1 to initialize the scope controls you will be using. *Initialize* means to set in standard positions. These standard settings are necessary so that, as you follow the directions on these pages, you'll see the same display on your scope crt as the displays pictured and described here.

Exercise 1. INITIALIZING THE SCOPE

Use the illustration at the end of this appendix to locate the controls described here.

1. DISPLAY SYSTEM CONTROLS: Set the INTENSITY control at midrange (about halfway from either stop). Turn the FOCUS knob completely clockwise.

2. VERTICAL SYSTEM CONTROLS: Turn the channel 1 POSITION control completely counterclockwise. Make sure the left VERTICAL MODE switch is set to CH 1. Move both VOLTS/DIV switches to the least sensitive setting (5V) by rotating them completely counterclockwise. Verify that both center CAL controls are locked in their detents (completely clockwise) and are pushed in (X1 vertical magnification). Set both input coupling switches (located below the VOLTS/DIV switches) to GND.

3. HORIZONTAL SYSTEM CONTROLS: Set the HORIZONTAL MODE switch to X1. Rotate the SEC/DIV switch to 0.5 ms (0.5 millisecond per division). Make sure the variable (CAL) control in the center of the SEC/DIV switch is locked in its detent (fully clockwise).

4. TRIGGER SYSTEM CONTROLS: Make sure the HOLDOFF control is rotated in MIN (completely counterclockwise). Set the TRIGGER MODE switch to P-P AUTO. Move the left SOURCE switch to CH 1.

5. Finally, verify that the LINE VOLTAGE SELECTOR switch on the rear panel is set for the proper nominal voltage source and that the proper fuse is installed. Then plug the scope into a properly grounded outlet and turn it on by pressing in the POWER pushbutton. Let the scope warm up for about five minutes—then you're ready to go!

The Display System

The oscilloscope draws a graph by moving an electron beam across a phosphor coating on the inside of the crt. The result is a glow, which, for a short time afterwards, traces the path of the beam. A grid of lines etched on the inside of the faceplate serves as the reference for measurements; this is the graticule shown in Fig. A.2.

Common controls for display systems include intensity and focus; some oscilloscopes also have a beam finder and trace rotation controls. At the top of the 2225, directly to the right of the crt, is the intensity control, labeled INTENSITY. Then come the beam finder (BEAM FIND), FOCUS, and TRACE ROTATION controls. These controls are

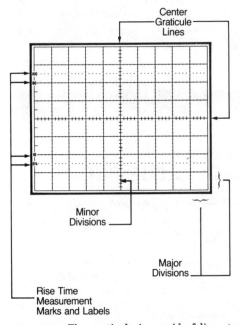

Figure A.2 The graticule is a grid of lines typically etched or silk-screened on the inside of the crt faceplate. Putting the graticule inside—on the same plane as the trace drawn by the electron beam—eliminates measurement inaccuracies called *parallax errors*. Parallax error occurs when the trace and the graticule are on different planes and the observer shifts slightly from the direct line of sight. Though different-sized crt's may be used, graticules are usually laid out in an 8-by-10 pattern. Each of the 11 vertical and 9 horizontal lines block off major divisions (also simply called *divisions*) of the screen. Labeling on the scope controls always refers to major divisions. The tick marks on each of the graticule lines represent minor divisions or subdivisions. Since scopes are often used for rise time measurements, 2200-Series scope graticules include special markings to aid in making rise-time measurements. There are dashed lines for 0 percent and 100 percent levels and labeled graticule lines for the 10 percent and 90 percent points (where rise time is measured).

described next, and their locations on the Tektronix 2225 are shown on the illustration at the end of this appendix.

Intensity. The INTENSITY control adjusts the brightness of the trace. It is necessary because you use a scope in different ambient-light conditions and with many kinds of signals. For instance, on square waves, because the slower horizontal segments look brighter than the faster vertical segments, you will probably want to turn up the intensity to make the fainter parts of the waveform easier to see.

The INTENSITY control is also useful because the intensity of a trace depends on two factors: how bright the beam is and how long it is on screen. As you select different sweep speeds (a sweep is one movement of the electron beam across the scope screen) with the SEC/DIV switch, the beam on and beam off times change—that is, the beam has either more or less time to excite the phosphor.

Beam finder. For convenience, the beam finder lets you locate the electron beam when it is off-screen. When you push the BEAM FIND button, you reduce the vertical and horizontal deflection voltages (more about deflection voltages later) and override the INTENSITY control so that the beam always appears within the 80-by-100 millimeter screen. When you see the quadrant of the screen in which the beam appears, you'll know which way to turn the HORIZONTAL and VERTICAL POSITION controls to reposition the trace. If you inadvertantly turned down the display intensity, you will still locate the trace, since intensity is automatically increased while BEAM FIND is pressed.

Focus. The electron beam of the scope is focused on the crt faceplate by an electrical grid within the tube. The FOCUS control adjusts that grid for optimum trace focus. On a 2200-Series scope, the focus circuit maintains focus settings over most intensities and sweep speeds.

Trace rotation. Another display control on the front panel of a 2200-Series instrument is TRACE ROTATION. This adjustment lets you electrically align the horizontal deflection of the trace with the fixed crt graticule. To avoid accidental misalignments when the scope is in use, the control is recessed and must be adjusted with an adjustment tool or a screwdriver.

If this seems like a calibration item that should be adjusted once and then forgotten, it is—for most oscilloscope applications. But the earth's magnetic field affects trace alignment, and when a scope is used in many different positions—as a service scope will be—it's very handy to have a front-panel trace-rotation adjustment.

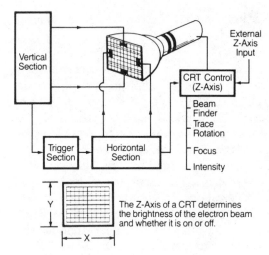

Figure A.3 The display system consists of the Z-axis, cathode-ray tube, and controls. To draw the graph of your measurements, the vertical system takes the input signal and supplies the Y-axis coordinates and the horizontal system supplies the X-axis coordinates. The trigger system takes input-signal information from the vertical system and uses it to start the horizontal sweep. The horizontal system also controls the Z-axis, which determines whether the electron beam (intensity) is turned on or *blanked* (off).

Using the display controls. The display system and its controls are shown as functional blocks in Fig. A.3. Use Exercise 2 to review the display controls.

Exercise 2. DISPLAY SYSTEM CONTROLS

In Exercise 1 you initialized your scope and turned on the power. Now find the display system controls labeled on the illustration at the end of Appendix A and use them as you follow these instruction.

1. BEAM FIND: Locate the position of the electron beam by pushing and holding in the BEAM FIND button; then use the channel 1 VERTICAL POSITION knob to position the trace on the center horizontal graticule line. Keep BEAM FIND pressed in and use the HORIZONTAL POSITION control to center the trace. Then release the beam finder.

2. FOCUS: The trace you have on the screen now should be out of focus. Make it as sharp as possible using the FOCUS control.

3. INTENSITY: Set the brightness so that you get a good display.

4. VERTICAL POSITION: Now use the VERTICAL POSITION control to line up the trace with the center horizontal graticule line.

5. TRACE ROTATION: Use a small screwdriver to rotate the TRACE ROTATION control in both directions. Notice that the trace appears to revolve clockwise and counterclockwise. When you finish, align the trace so that it is parallel to the horizontal graticule line closest to it. After aligning the trace, you may have to use the VERTICAL POSITION control again to set the trace on the graticule line.

You have now used all the scope's display-system controls. If, at the end of a chapter you don't plan to continue right away, be sure to turn off your scope.

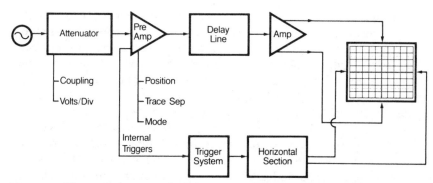

Figure A.4   The vertical system of a Tektronix 2200-Series scope consists of two identical channels, though only one is shown in the drawing. Each channel has circuitry to couple an input signal to that channel, attenuate the signal (that is, reduce it) preamplify it, delay it, and finally amplify the signal for use by the display system. The delay line lets you see the beginning (or leading edge) of a waveform even when the scope is triggering on it.

The Vertical System

Your scope's vertical system supplies the display system with the vertical—or Y-axis—information for the graph on the crt screen. The vertical system takes the input signals, develops *deflection voltages,* then uses the deflection voltages to control—that is, deflect—the crt electron beam.

The vertical system also gives you a choice of how you connect the input signals (called *coupling,* described below). And it provides internal signals for the trigger circuit, which is described in a later section. Figure A.4 schematically illustrates the vertical system.

The set of dual vertical-system controls—see the illustration at the end of this appendix for their locations—include the POSITION controls, the VOLTS/DIV sensitivity controls, and the input coupling switches. Because the 2225 has two channels, there are one set of these switches and a POSITION control for each channel. There are also a trace-separation control (TRACE SEP) and three mode switches. The left MODE switch lets you select channel 1, channel 2, or both signals for display; the center switch lets you select either an uninverted or inverted channel 2 display; and the right MODE switch lets you select a display that algebraically combines channel 1 and channel 2 (ADD), an alternated display (ALT), or a chopped display (CHOP).

For the exercises in this section, you'll need a 10X probe like the Tektronix P6103 10X Probe supplied with every 2225 Oscilloscope.

Vertical position.   The vertical POSITION controls let you place the trace exactly where you want it on the screen. The position control for each channel changes the vertical placement of its respective trace.

Input coupling. The input coupling switch for each vertical channel lets you control how the input signal is coupled to the vertical channel. When DC (the abbreviation stands for *direct current*) input coupling is selected, you see all of an input signal. With AC (*alternating current*) coupling, the constant (static) signal components are blocked and only the alternating (dynamic) components of the input signal reach the channel. For an illustration of the differences, see Fig. A.5.

The middle position of the coupling switches is marked GND, for *ground*. This setting disconnects the input signal from the vertical system. And when P-P AUTO triggering mode is set, the display shows a baseline trace indicating chassis ground or zero volts. The position of the trace on the screen in this mode is the ground reference

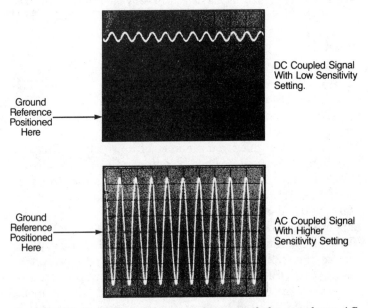

Ground Reference Positioned Here

DC Coupled Signal With Low Sensitivity Setting.

Ground Reference Positioned Here

AC Coupled Signal With Higher Sensitivity Setting

Figure A.5 Vertical channel input coupling controls let you choose AC and DC input coupling and ground (GND). Using DC coupling connects the entire input signal to the vertical channel. AC coupling blocks constant (static) signal components and only connects alternating (dynamic) components to the vertical channel. The GND position disconnects the input signal from the vertical system. And when P-P AUTO triggering mode is set, the display is a baseline trace indicating chassis ground or zero volts. The position of the trace on the screen in this mode is the ground reference level. Switching from DC to GND and back is a handy way of measuring signal voltage levels with respect to chassis ground. Using the GND position does not ground the signal in the circuit you're probing. The AC coupling setting is handy when the entire signal (alternating plus constant components) is too large for the VOLTS/DIV settings, as shown in the top photo. Eliminating the constant component lets you look at the alternating signal with a more convenient VOLTS/DIV setting, as shown in the bottom photo.

level. Switching from DC to GND and back is a handy way of measuring signal voltage levels with respect to chassis ground. Using the GND position does not ground the signal in the circuit you're probing.

Vertical sensitivity. The VOLTS/DIV switch determines the sensitivity of each vertical channel. The ability to select a range of sensitivities extends the depth of possible applications. A multipurpose scope can accurately display a broad range of signal levels, from millivolts to many volts. With a maximum sensitivity of 500 µV (0.5 mV) per division, the Tektronix 2225 Oscilloscope is four to ten times more sensitive than other comparable instruments.

The slash (/) in VOLTS/DIV stands for *per*, and each setting of this switch is a scale factor. For example, a setting of 20 mV means 20 millivolts per division. Using the VOLTS/DIV switch to change sensitivity also changes the *scale factor*, which is the value of each major division on the screen. Each setting of the switch is marked with a number that represents the scale factor for that channel. To illustrate, suppose the VOLTS/DIV setting is 5 V; then each of the eight vertical major divisions represents 5 volts, and the entire screen can show 40 volts from bottom to top. With a setting of 500 µV per division, the screen can display 4 millivolts from top to bottom.

Vertical magnification. The 500 µV per division sensitivity of the 2225 is achieved by using an additional gain (or amplifier) stage. It is activated by pulling out the VOLTS/DIV variable (CAL) knob and it increases the selected sensitivity by a factor of 10. Each channel has its own independent amplifier that operates with all VOLTS/DIV settings. When the X10 vertical magnification feature is selected, bandwidth is reduced to 5 MHz. Thus it can be used as bandwidth-limiting mode when high-frequency noise is present on the input signal.

Probe scale factor. The probe you use influences the scale factor. On the front panel, next to each VOLTS/DIV switch, notice the two bracketed areas labeled 1X and 10X PROBE. The right area shows the scale factor for a standard 10X probe. The left area shows the scale factor for a 1X probe.

Variable VOLTS/DIV. The CAL control in the center of each VOLTS/DIV switch provides a continuously variable change in the scale factor—up to more than 2.5 times the VOLTS/DIV setting. This control is used most commonly to align waveforms vertically with the 0 percent and 100 percent graticule lines when making risetime measurements. The variable sensitivity control is also useful for making quick amplitude comparisons of a series of signals. You could, for example, take a known signal of almost any amplitude and use the CAL control to make sure the waveform fits exactly between major division graticule lines. Then, using the same vertical channel to look at other signals, you could quickly see whether later signals had the same amplitude.

Channel 2 inversion.   For differential measurements (described in Part II) you can invert the polarity of the channel 2 signal by using the middle VERTICAL MODE switch. When the switch is set to the right (CH 2 INVERT), the channel 2 signal is inverted; when it is set to the left (NORM), both channels have the same polarity.

Vertical operating modes.   Scopes become more useful as the number of vertical display modes increases. You have several display choices with your Tektronix 2225 Oscilloscope. These modes are controlled by the left and right VERTICAL MODE switches. With these two switches, you can select channel 1 alone, channel 2 alone, both channels in either alternated or chopped mode, and both channels algebraically summed.

To display only channel 1, set the left switch to CH 1; to display only channel 2, set it to CH 2.

To see both channels in the alternated mode, set the left switch to BOTH and the right switch to ALT. This causes each channel to be drawn alternately—the scope completes first one sweep on channel 1, then one sweep on channel 2, and then repeats. This mode is used with medium to fast signals—when the SEC/DIV switch is set to 0.5 ms or faster.

To display both channels in a chopped mode, set the left switch to BOTH and the right switch to CHOP. Now the scope will draw small parts of each signal by switching back and forth between them. The switching rate is so fast that your eyes naturally fill in the gaps, and the waveform looks whole. This mode is typically used with slow signals requiring sweep speeds of 1 ms per division and slower.

Both chopped and alternated modes are provided so that you can look at two signals at any sweep speed. The ALT mode first draws one trace then the other, not both at the same time. This works well at faster sweep speeds, when your eyes can't detect the alternating action. To see two signals (simultaneously) at slower sweep speeds, use the CHOP mode.

To see the two input signals displayed as one waveform, set the left switch to BOTH and the right switch to ADD. This gives you an algebraically combined signal—either channel 1 + channel 2, or differentially combined (channel 1 − channel 2), when channel 2 is inverted.

Sweep separation.   The 2225 alternate-magnification scope also has a sweep separation control. Labelled TRACE SEP, it's used to change the vertical position of a horizontally magnified trace with respect to the unmagnified trace of the same signal. Using the TRACE SEP control in conjunction with the vertical POSITION controls lets you place all four traces (the magnified and unmagnified traces for each of two channels) on the screen so that they don't overlap. Horizontal magnification is discussed in the next section.

Using the vertical controls. Before using the vertical system controls, make sure all the front-panel controls are positioned where you left them at the end of the last chapter:

- Display

  INTENSITY and FOCUS:
  Bright, crisp trace

- Vertical

  POSITION (Channel 1): Fully counterclockwise
  MODE: CH 1
  VOLTS/DIV (both): 50 V (10X PROBE)
  VOLTS/DIV Variable (both): CAL detent (fully clockwise) and pressed in (no vertical magnification)
  Input Coupling (both): GND

- Horizontal

  MODE: X1
  SEC/DIV: 0.5 ms
  SEC/DIV Variable: CAL detent (fully clockwise)

- Trigger

  MODE: P-P AUTO
  HOLDOFF: MIN (fully counter clockwise)
  SOURCE: CH1

Now connect your 10X probe to the CH 1 BNC connector on the front panel of your scope. (BNC means *bayonet Neill-Concelman* and is named for Paul Neill, who developed the N-Series connectors at Bell Labs, and Carl Concelman, who developed the C-Series connectors.)

Attach the tip of the probe to the PROBE ADJUST terminal. Probes come with an alligator-clip ground strap grounding the probe to the circuit under test. Clip the ground lead onto the collar of the CH 2 BNC connector as shown in Fig. A.6.

Using the callouts on the figure at the back of Appendix A to remind yourself of control locations, follow the directions in Exercise 3 to review the vertical system controls.

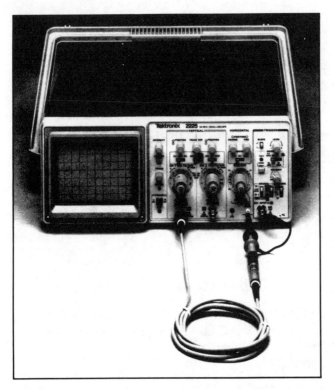

Figure A.6 The Tektronix P6103 10X probe connects to the BNC connector of either channel 1 (shown) or channel 2. Although not shown in the photo, the probe's ground strap is usually connected to the ground of the circuit you are working on. The PROBE ADJUST terminal is located at the bottom of the HORIZONTAL section on the front panel.

Exercise 3. VERTICAL SYSTEM CONTROLS

Compensating Your Probe

1. Turn on the scope and turn the CH 1 VOLTS/DIV switch clockwise to 0.5 V. The P6103 is a X10 probe, so use the VOLTS/DIV readout to the right (labeled 10X PROBE).

2. Switch channel 1 input coupling to AC.

3. If the signal on the screen isn't steady, turn the trigger LEVEL control until the signal stops moving and the triggered light (labeled TRIG'D) is illumined. If needed, use the FOCUS and INTENSITY controls to adjust signal sharpness and brightness.

4. Next, compensate the probe using the screwdriver adjustment in the probe head. To expose the adjustment screw, rotate the sleeve on the probe head until you see the low-frequency compensating (LF COMP) capacitor. Then, while observing the screen display, use a plastic adjustment tool to turn the LF COMP capacitor until the tops and bottoms of the square wave are flat (see photograph on page 272). You'll find more information about probes starting on page 263.

Controlling Vertical Sensitivity

1. The probe adjustment signal is an approximately 1-kHz square wave with about 0.5 V amplitude. The scale factor for channel 1 is currently set at a half-volt per division. At this setting every major division on the screen represents half a volt. Use the channel 1 POSITION control to line up the bottom edge of the waveform with the center horizontal graticule line. Notice that the top of the square wave is just touching the next major division line, which demonstrates that the probe-adjustment signal is indeed about 0.5 V.

2. Turn the VOLTS/DIV switch clockwise two more stops. The channel 1 scale factor is now 0.1 volts per division (with a 10X probe), and the signal—still 0.5 V—is now about five major divisions in amplitude.

3. Turn the VOLTS/DIV variable (CAL) control counterclockwise out of its detent and notice what happens. When you've turned it fully counterclockwise, the signal should be less than two major divisions in amplitude, because the scale factor was reduced by at least 2.5 times. Now return the variable (CAL) control to its detent (fully clockwise).

4. Set the channel 1 VOLTS/DIV switch to 5 V and pull out (towards you) the CAL-X10 knob. Observe that the displayed waveform is one division in amplitude, indicating that vertical sensitivity is now back to 0.5 V per division. Push in the CAL X10 knob and return the CH 1 VOLTS/DIV switch to 0.1 V.

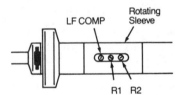

Location of compensation adjustments
on the Tektronix P6103 probe.

Coupling the Signal

1. Switch the channel 1 input coupling to GND and position the trace on the center horizontal graticule line. Switch back to AC coupling and note that the waveform remains centered on the screen. Move the CH 1 VOLTS/DIV switch to 0.5 V and notice that the waveform is still centered around the zero reference line.

2. Switch input coupling to DC. The top of the probe adjustment signal should be on the center graticule line, and the signal should reach to the next lower major division. Now you can see the difference between AC and DC coupling. The AC coupling blocks the constant part of the signal and just shows you a one half volt, peak-to-peak square wave centered on the zero reference that you set at the center of the screen. The DC coupling shows you that the constant component of the square wave is negative-going with respect to ground, because with DC coupling all signal components are connected to the vertical channel.

(Continued)

The Vertical Mode Controls

1. So far you've been using the scope to see what channel 1 can tell you, but that's only one of many possible vertical modes. Look at the trace for channel 2 by setting the left VERTICAL MODE switch to CH 2. Input coupling for channel 2 should still be at GND, so what you'll see is the ground reference line. Use the channel 2 POSITION control to line up this trace with the second graticule line from the top of the screen.

2. Now move the left VERTICAL MODE switch to BOTH. This lets you pick one of the vertical modes controlled by the right MODE switch. Move the right switch to ALT, the alternated mode. In ALT the trace for one channel is drawn completely before the scope switches to draw the trace for the other channel. You can see this happening when you slow down the sweep speed, so turn the SEC/DIV switch counterclockwise to 0.1 second per division. Now notice that the two dots from channel 1, which is AC-coupled, move across the screen for one sweep, then the single dot from channel 2 moves across the screen.

3. Turn the SEC/DIV switch back to 1 ms and switch to CHOP as your vertical mode. The display looks a lot like the alternated mode, but it is achieved in an entirely different way. In ALT you saw that one channel's signal was completely written before the other one was started. But when you're looking at slow signals, having only one trace at a time on screen can be a bother. In CHOP the scope switches back and forth very quickly between the two traces so that a part of one signal is drawn, then a part of the other, then the process repeats. When you look at the screen, both signals seem continuous because the scope is chopping back and forth at a very fast rate—approximately 500 kHz in the 2200-Series scopes. The CHOP mode is most useful for slow sweep speeds, and ALT is best for faster sweeps.

4. Turn the SEC/DIV switch back to 0.5 ms. There's one more vertical mode—ADD. In this mode, the two signals are either algebraically combined (channel 1 + channel 2) or differentially combined when channel 2 is inverted (channel 1 − channel 2). To see this mode in operation, set the VERTICAL MODE switches to CH 2-NORM-ADD. Connect a probe to the CH 2 BNC connector and get a display for channel 2 just like you did for channel 1. Then set the left MODE switch to BOTH. Now you can see the combined signal roughly halfway between where the two separate signals were.

The Horizontal System

To draw a graph, your scope needs both horizontal and vertical data. The horizontal system supplies the second dimension by providing deflection voltages that move the electron beam horizontally. It also contains a sweep generator that produces a sawtooth waveform, or ramp (see Fig. A.7). The ramp is used to control the scope's sweep rates.

The sweep generator makes possible the unique functions found in today's modern oscilloscope. The circuit that produces a linear rate of rise in the ramp—a refinement pioneered by Tektronix—was one of the most important advances in oscillography. It meant that horizontal beam movement could be calibrated directly in units of time, and this gives you the ability to measure time between events much more accurately.

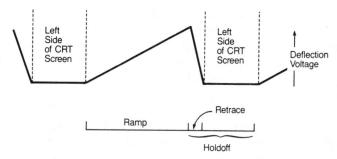

Figure A.7   The sawtooth waveform is a voltage ramp produced by the sweep generator. The rising portion of the waveform is called the *ramp*, the falling edge is called the *retrace,* and the time between ramps is called the *holdoff*. The sweep of the electron beam across the screen is controlled by the ramp, and the return of the beam to the left side of the screen takes place during the retrace. During the holdoff time, the electron beam remains on the left side of the screen before starting the sweep.

Because the sweep generator is calibrated in time, it is usually called the *time base.* The time base lets you observe the signal for the unit of time you select—either for very short times, measured in nanoseconds or microseconds—or for relatively long times of several seconds.

The horizontal system controls on the Tektronix 2225 Oscilloscope are shown in the illustration at the back of this appendix; the COARSE and FINE POSITION controls are near the top of the panel, and the MODE control is below them. Variable sweep-speed control is afforded by the CAL knob in the center of the SEC/DIV switch. At the bottom of this column of controls is the magnification (MAG) control.

Horizontal operating modes.   The 2225 is capable of highly accurate timing measurements and a wide range of time-delay measurements, even though it is a single-time-base scope. Single-time-base scopes usually have only one operating mode, but the 2225 offers three. Through the interaction of the MODE and MAG switches, you not only can display the unmagnified trace (1X) alone, but also can display the same trace magnified 5, 10, or 50 times—either alone (MODE set to MAG) or together with the unmagnified trace (MODE set to ALT with the MAG switch set to either X5, X10, or X50).

Oscilloscopes such as the Tektronix 2235 have two time bases, which, in addition to waveform expansion, allow the user to make very accurate (1 percent) differential timing measurements. These are accomplished by setting the two time bases to different SEC/DIV values and using a calibrated delay dial.

The 2225 offers the best mix of measurement capabilities between single-time-base scopes and dual-time-base scopes. Its X1 MODE provides the same display as a basic single-time-base scope and as the A sweep on a dual-time-base scope. With the 2225 HORIZONTAL MODE set to ALT (alternate), the display alternates between the unmagnified trace (X1) and the magnified trace (X5, X10, or X50). This is the same as a dual-time-base scope operating in an alternate mode and displaying both the A time base sweep and the B time-base sweep. When the 2225 is set to MAG MODE, only the magnified trace is displayed. This is the same as a dual-time-base scope that is set to display only the B sweep.

The advantage of the 2225 is that it has most of the capabilities of a dual-time-base scope, yet it offers the operational simplicity of a single-time-base scope.

Sweep speeds. The SEC/DIV switch lets you select the rate at which the beam sweeps across the screen; changing SEC/DIV switch settings let you look at longer or shorter time intervals of the input signal. Like the vertical system VOLTS/DIV switch, the control markings refer to the horizontal scale factors applied to the on-screen trace. If the SEC/DIV setting is 1 ms, each horizontal major division represents 1 ms, and the total screen will be 10 ms wide.

All the instruments in the Tektronix 2200-Series offer sweep speeds from one-half second to 0.05 μs per division. Markings that appear on the SEC/DIV switch are:

0.5 s	0.5 second
0.2 s	0.2 second
0.1 s	0.1 second
50 ms	50 milliseconds (0.05 second)
20 ms	20 milliseconds (0.02 second)
10 ms	10 milliseconds (0.01 second)
5 ms	5 milliseconds (0.005 second)
2 ms	2 milliseconds (0.002 second)
1 ms	1 milliseconds (0.001 second)
0.5 ms	0.5 milliseconds (0.0005 second)
0.2 ms	0.2 milliseconds (0.0002 second)
0.1 ms	0.1 milliseconds (0.0001 second)
50 μs	50 microseconds (0.00005 second)
20 μs	20 microseconds (0.00002 second)
10 μs	10 microseconds (0.00001 second)
5 μs	5 microseconds (0.000005 second)
2 μs	2 microseconds (0.000002 second)

1 μs	1 microseconds (0.000001 second)
0.5 μs	0.5 microseconds (0.0000005 second)
0.2 μs	0.2 microseconds (0.0000002 second)
0.1 μs	0.1 microseconds (0.0000001 second)
0.05 μs	0.05 microseconds (0.00000005 second)

Scopes also have an X-Y setting often placed on the SEC/DIV switch, for making the X-Y measurements described on page 280.

Variable SEC/DIV.   Besides the calibrated speeds, you can change any sweep speed by turning the CAL control (in the center of the SEC/DIV switch) counterclockwise. This control slows the sweep speed by at least 2.5:1, making the slowest sweep 1.25 seconds per division (0.5 seconds x 2.5). Remember that the calibrated position is completely clockwise (in detent).

Horizontal magnification and alternate magnification.   When you select either ALT or MAG, you are multiplying the currently set sweep speed by the chosen magnification factor. For example, if the SEC/DIV switch setting is 0.05 μs and the MAG switch setting is X5, then the magnified sweep speed is a fast 10 ns per division; and when MAG is X10, the magnified sweep is a very fast 5 ns per division. Magnification is useful when you want to look at signals and see details that occur very closely together. It is especially helpful for measuring digital signal time.

Expanding a waveform for measurement using the alternate magnification feature is accomplished with the HORIZONTAL MODE switch (set to ALT) and the POSITION control. First locate a point of interest on the unmagnified (X1) trace.

Then, simply position that point horizontally to the center vertical graticule line. The magnified trace will be positioned to display the selected point at or about the center vertical graticule line, which is called the magnifier registration point. Use the TRACE SEP control to position the magnified trace either above or below the unmagnified trace. It's as simple as that.

Horizontal position.   You use the HORIZONTAL POSITION controls just as you use the VERTICAL POSITION controls—to change the location of waveforms on the screen. The 2225 has two HORIZONTAL POSITION controls—COARSE and FINE. The FINE control is well suited for adjusting the magnified sweep. As previously explained, the HORIZONTAL POSITION controls are also used in conjunction with the alternate magnification feature.

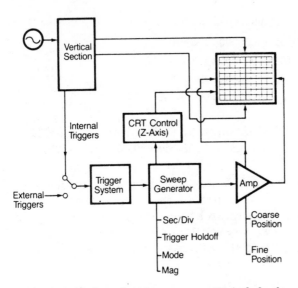

Figure A.8 Horizontal system components include the sweep generator and the horizontal amplifier. The sweep generator produces a sawtooth waveform that is processed by the amplifier and applied to the horizontal deflection plates of the crt. The horizontal system also provides the scope's Z-axis, which determines whether the electron beam (intensity) is turned on or turned off (*blanked*).

Using the horizontal controls.   As you can see in Fig. A.8, the horizontal system can be divided into two functional blocks, the horizontal amplifier and the sweep generator.

To familiarize yourself with the horizontal system controls, follow the directions in Exercise 4 and refer to the illustration at the back of this appendix for control locations. First, make sure the front-panel controls have these settings:

VERTICAL MODE: CH 1 - NORM

CH 1 VOLTS/DIV: 0.5 V (10X PROBE)

HORIZONTAL MODE: X1

SEC/DIV: 0.5 ms

TRIGGER MODE: P-P AUTO

TRIGGER SOURCE: CH 1

Exercise 4. HORIZONTAL SYSTEM CONTROLS

1. Be sure your probe is connected to the CH 1 input BNC and its tip is attached to the PROBE ADJUST terminal. Turn on your scope, set the channel 1 input coupling switch to GND, and align the trace along the center horizontal graticule line using the channel 1 POSITION control. Then switch to AC input coupling.

2. Now you can use the horizontal system of your scope to look at the probe-adjustment signal. Move the waveform with the HORIZONTAL COARSE POSITION control until one rising edge is lined up with the center vertical graticule line. Examine the screen to see where the leading edge of the next pulse crosses the horizontal centerline of the graticule. Count the major and minor graticule markings along the center horizontal graticule between the successive pulse leading edges. Note the number.

3. Change sweep speed to 0.2 ms and line up a rising edge with the vertical graticule line on the left edge of the screen. Again count divisions to the next rising edge. Because the switch was changed from 0.5 ms to 0.2 ms, the waveform will look two and one-half times as long as before. The signal hasn't changed, of course, only the scale factor.

4. You'll find the variable control in the middle of the SEC/DIV switch. It's labeled CAL. The SEC/DIV switch settings are calibrated when the CAL control is fully counterclockwise in its detent. Move this control from its detent and rotate it fully counterclockwise to see the effect on sweep speed. Notice that now the cycles of the waveform are at least two and one-half times smaller. Return the CAL control to its detent.

5. Turn the SEC/DIV switch back to 0.5 ms, set the HORIZONTAL MODE switch to ALT, and set the MAG switch to X10. Notice that a second trace appears on the screen. This second trace is a 10X magnification of the original trace; that is, its sweep speed is ten times faster. For example, the sweep speed of the magnified trace now is 0.05 ms per division, while the sweep speed of the original (unmagnified) trace remains 0.5 ms.

6. The separation between the magnified and unmagnified traces can be adjusted with the TRACE SEP control. Use this control to set a convenient spacing between the two traces. Notice that only the magnified trace moves. You may have to use the VERTICAL POSITION control to center the display again.

7. While your scope is magnifying the probe adjustment signal, use the HORIZONTAL COARSE POSITION control. Its range is now magnified as well, and the combination of the magnified signal and COARSE POSITION control rotation lets you examine small parts of a waveform in great detail. At this point, try using the FINE POSITION control with the X5 and X50 magnification levels. Notice the effects.

8. Now switch HORIZONTAL MODE to MAG and notice that only the magnified trace remains on the display. Switch between X5, X10, and X50 magnification levels and notice the effect.

9. Return your scope to its normal sweep speed range and display by setting the MODE switch to X1.

The Trigger System

So far you've found that the display system draws the waveforms on the screen, the vertical system supplies the vertical information for the graph, and the horizontal system provides the time axis. In other words, you know how the oscilloscope draws a graph. The only thing missing is the "when." How does the scope "know" when to start drawing the signal?

The trigger is the "when." It determines the instant that the scope starts drawing the signal. Triggering is important because acquiring time-related information is one of the primary reasons for using a scope. Equally important is that the graph of each signal should start predictably at the same point of the waveform so that the display remains stable.

The graph you see on the screen isn't a static one even though it appears to be. It's always changing—being updated with the input signal. If you're using the 0.05 μs SEC/DIV setting, the scope is drawing one graph every 0.5 μs (0.05 μs per division times 10 screen divisions). That's 2,000,000 graphs every minute (not counting retrace and holdoff times, which we'll get to shortly). Imagine the jumble on the screen if each sweep started at a different place on the signal.

But each sweep does start at the right time—if you make the right trigger system control settings. Here's how it's done. You use the TRIGGER SOURCE switches to tell the trigger circuit which trigger signal to select—channel 1, channel 2, line voltage, or an external signal. To use an external trigger signal, you connect it to the trigger system circuit using the external input (EXT INPUT) connector. You use the COUPLING switch to determine how the selected signal is to be coupled to the trigger circuits. Next you set the SLOPE and LEVEL controls to tell the trigger circuit to recognize a particular voltage level on the trigger signal. Then every time that level occurs, the sweep generator is turned on. See Fig. A.9 for a diagram of the process.

Instruments like those in the Tektronix 2200-Series offer a variety of trigger controls. Besides those already mentioned, you also have controls that determine how the trigger system operates (*trigger operating mode*) and how long the scope waits between triggers (*holdoff*).

Trigger control locations are illustrated in the photograph at the end of this appendix. All are on the far right of the front panel. The SLOPE and LEVEL controls are grouped at the top; below them is the trigger MODE switch and below that is the HOLDOFF control. A set of three switches controls the trigger SOURCE and trigger COUPLING. At the bottom is the external trigger input (EXT INPUT) connector.

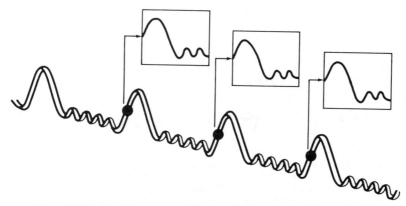

Figure A.9 Triggering gives you a stable display because the same trigger point starts the sweep each time. The SLOPE and LEVEL controls define the trigger points on the trigger signal. When you look at a waveform on the screen, you're seeing all those sweeps overlaid in what appears to be a single picture.

Trigger level and slope. Controls for these functions define the trigger point. The SLOPE control determines whether the trigger point is found on the rising or the falling edge of a signal. The LEVEL control determines where on that edge the trigger point occurs (see Fig. A.10).

Variable trigger holdoff. Not every trigger event can be accepted as a trigger. The trigger system recognizes only one trigger during the sweep interval. Also, it does not recognize a trigger during retrace and for a short time afterward (the holdoff period). The retrace, as you remember from the last chapter, is the time the electron beam takes to return to the left side of the screen to start another sweep. The holdoff period provides additional time beyond the retrace and is used to ensure that your display is stable, as illustrated in Fig. A.11.

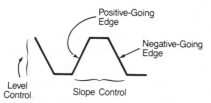

Figure A.10 Slope and level controls determine where on the trigger signal the trigger actually occurs. The slope control specifies either a positive (also called the *rising* or *positive-going*) edge or a negative (*falling* or *negative-going*) edge. The level control lets you choose where on the selected edge the trigger event will occur.

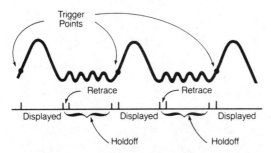

Figure A.11 Trigger holdoff time ensures valid triggering. In the drawing, only the labeled points start the display because no other trigger can be recognized after the sweep starts or during the retrace and holdoff periods. The retrace and holdoff times are necessary because the electron beam must be returned to the left side of the screen when the sweep ends, and because the sweep generator needs the reset time. The crt Z-axis in *blanked* (or turned off) between sweeps and *unblanked* during sweeps.

Sometimes the normal holdoff period isn't long enough to ensure that you get a stable display; this possibility exists when the trigger signal is a complex waveform with many possible trigger points on it. Though the waveform is repetitive, a simple trigger might get you a series of patterns on the screen instead of the same pattern each time. Digital pulse trains are a good example; each pulse is very much like any other, so there are many possible trigger points, not all of which result in the same display.

You need a way to control when the scope accepts a trigger point. The variable TRIGGER HOLDOFF control provides this capability. Because it adjusts the holdoff time of the sweep generator, this control is part of the horizontal system; but its function interacts with the trigger controls. Figure A.12 depicts a situation in which the variable holdoff function is useful for extending the holdoff time.

Trigger sources. Trigger sources are grouped into two categories that depend on whether the trigger signal is provided internally or externally. The source makes no difference in how the trigger circuit operates, but the internal triggering usually uses the signal that is being displayed. That has the advantage of letting you see where you're triggering.

The two TRIGGER SOURCE switches on the front panel determine the source signal for the trigger. Internal triggering sources are enabled when you move the left SOURCE switch to the appropriate channel setting (CH 1 or CH 2), which allows you to trigger the scope

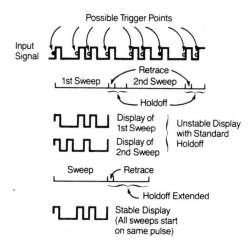

Figure A.12 The variable hold-off control causes the scope to ignore some potential triggering points. In the example, all the possible trigger points on the input signal would result in an unstable display if they were accepted as sweep triggers. Changing the holdoff time to establish a trigger point that appears on the same pulse in each repetition of the input signal is the only way to ensure a stable waveform.

on the signal coming from the selected channel. When triggering on one channel, you set the scope to trigger the sweep on some part of the waveform present on that channel.

You can also set the TRIGGER COUPLING switch to VERT MODE. When this position is selected, the scope's VERTICAL MODE switches determine what signal is used for triggering. If the VERTICAL MODE switch is set to CH 1, then the signal on channel 1 triggers the scope. If the switch is set to CH 2, then the channel 2 signal triggers the scope.

If the switch is set to BOTH-ALT, then both signals alternately trigger the scope. This is accomplished by alternating the channel 1 and channel 2 trigger signals synchronously with the display. In the BOTH-ADD mode, the algebraic sum of channel 1 and channel 2 is the triggering signal. And in the BOTH-CHOP mode, the scope triggers as it does in ADD mode, which prevents the instrument from triggering on the chop frequency instead of on your signals.

As you can see, VERT MODE triggering is a kind of automatic source selection that you can use when switching back and forth between vertical modes to look at different signals. This mode makes possible alternate triggering, with the scope triggering first on one channel then on the other channel. That means you can look at two completely unrelated signals at the same time. Most scopes can only trigger on either one channel or the other when the two input signals are not synchronous.

But triggering on the displayed signal isn't always what you need, so external triggering is also provided. This can give you more control over the display. To use an external trigger, set the left SOURCE switch to EXT and the right SOURCE switch to either EXT/10 or

EXT. Connect the external triggering signal to the EXT INPUT connector on the front panel. External triggering often is useful in digital design and repair—when you might want to look at a long train of very similar pulses while triggering with an external clock or with a signal from another part of the circuit.

The LINE position on the source switch gives you another triggering possibility—the power line. Line triggering is useful any time you're looking at circuits dependent on the power-line frequency. Examples include devices like lighting equipment and power supplies.

The following table describes all the trigger sources possible on the Tektronix 2225 Oscilloscope.

Triggering operating modes. The 2225 Oscilloscope can operate in four trigger modes: normal, peak-to-peak automatic (includes television line), television field, and single sweep.

In the peak-to-peak automatic mode (P-P AUTO), a timer starts after a triggered sweep starts. If another trigger is not generated before the timer runs out, a second trigger is generated anyway, causing the bright baseline trace to appear—even when there is no signal applied to the channel input. Timer circuits are designed to run out if the trigger-signal frequency is less than 20 Hz. In the 2200-Series scopes, the peak-to-peak automatic mode is a peak-detecting mode. For most of the signals you'll be measuring, the peak-to-peak automatic mode matches the TRIGGER LEVEL control range to the trigger signal

TABLE 2225   Triggering Signals with Any Trigger Mode

Switch settings			
Trigger source			
Left	Right	Vertical Mode	Triggering signal
CH1		CH1	Channel 1 input
VERT MODE	Disabled	BOTH-ALT	Alternatives between Channel 1 and Channel 2 inputs. Each input signal triggers its own display.
		BOTH-CHOP	Algebraic sum of Channel 1 and Channel 2 inputs.
		BOTH-ADD	
CH2		CH2	Channel 2 input.
EXT	LINE	Any	Power line.
	EXT/10		Signal applied to EXT INPUT connector and attenuated by a factor of 10.
	EXT		Signal applied to EXT INPUT connector

peak-to-peak amplitude. Because of this automatic action, you will probably never set the TRIGGER LEVEL control outside the signal range. No matter where the LEVEL control is set, a triggered sweep will occur. The P-P AUTO mode lets you trigger on signals having a changing amplitude pattern or waveshape—and it lets you do this without adjusting the LEVEL control.

The normal trigger mode (labeled NORM on the TRIGGER MODE switch) is one of the most useful, because it can handle a wider range of triggering signals than any other mode—from dc to 50 MHz. It's used primarily for very low frequency signals (less than 100 Hz). When there's no trigger, the normal mode does not permit a trace to be drawn on the screen.

Another useful operating mode is television triggering. Most scopes with this mode let you trigger on TV fields at sweep speeds of 0.1 ms per division and slower and on TV lines at 50 μs per division or faster. With a 2200-Series scope, you can trigger on either TV fields or TV lines at any sweep speed. For field triggering, set the TRIGGER MODE switch to TV FIELD, and for line triggering, set it to P-P AUTO setting/TV LINE.

You'll probably use the normal and peak-to-peak automatic modes most often—NORM because it's the most versatile and P-P AUTO because it's essentially automatic. But you should be aware of a couple limitations in the P-P AUTO mode. It's possible to have a low-frequency signal with a repetition rate that is mismatched to the run-out of the automatic mode timer, thus causing the signal to be unsteady. Also, the P-P AUTO mode cannot trigger on very low frequency trigger signals. However, for these cases, use the NORM mode, which gives you a stable trigger at any frequency.

Single-sweep mode (labeled SGL SWP) operates exactly as its name implies—it triggers a sweep only one time. After selecting SGL SWP TRIGGER MODE, the trigger system must be readied, or armed, to accept the very next trigger event that occurs. This is done by momentarily pressing then releasing the RESET button, which also causes the READY-TRIG'D indicator to illuminate. With the trigger circuit armed, the crt screen is blank. When the trigger event or signal occurs, the sweep is started. And when one sweep is completely across the crt, the trigger system is halted. This stops any more sweeps from occurring and extinguishes the READY TRIG'D indicator.

The SGL SWP mode typically is used for waveform photography and for babysitting while looking for glitches. To perform the latter, set up the scope for displaying a single sweep, being careful that the SLOPE and LEVEL controls are both complementary to the event you are seeking. Then, if the scope user leaves the test area and later returns to see the READY-TRIG'D indicator extinguished, the operator knows

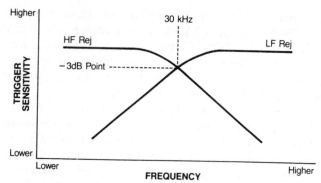

Figure A.13 Triggering stability can be improved with HF REJ and LF REJ COUPLING. The high- and low-frequency rejection filters roll off at 30 kHz.

that the event occurred. An oscilloscope camera could be used to capture this event on a permanent film record. This is achieved by leaving the shutter open, using the B (BULB) setting.

Trigger coupling. Just as you may select either alternating or direct coupling when connecting an input signal to the scope's vertical system, the 2225 Oscilloscope lets you choose the kind of coupling needed to connect either an internal or an external trigger signal to the trigger-system circuits.

Besides AC and DC coupling, the trigger system on the 2225 also has high-frequency rejection (HF REJ) and low-frequency rejection (LF REJ), which are useful for eliminating specific frequency components from the trigger signal. As its label implies, HF REJ removes the high-frequency portion of the triggering signal, allowing only the low-frequency components to pass on to the triggering system and subsequently start the sweep. Selecting LF REJ accomplishes just the opposite effect. These two frequency rejection features are useful for eliminating noise that may be riding on top of input signals. This noise may be preventing the trigger signal from starting the sweep at the same point every time. Typically, the HF and LF rejection filters roll off at 30 kHz, as illustrated in Fig. A.13.

Here's a review of the 2200-Series trigger modes:

Trigger operating mode	Trigger mode setting
Automatic peak-to-peak and television line	P-P AUTO
Normal	NORM
Television field	TV FIELD
Single sweep	SGL SWP

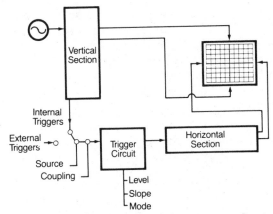

Figure A.14 The trigger circuit and its controls are shown here. Trigger SOURCE selects the signal that will be used for triggering the sweep. The COUPLING control affects the connection of the selected trigger signal to the trigger circuit. The LEVEL and SLOPE controls determine where the trigger point will be on the trigger signal. And the MODE control determines the operations of the trigger circuit.

Remember that for CH 1, CH 2, and VERT MODE TRIGGER SOURCE, the triggering signal first goes through the VERTICAL input coupling circuit. Thus, if AC input coupling is selected, only the AC component of the input signal gets to the trigger system—even if TRIGGER COUPLING is set to DC. The 2225's greater trigger versatility is outlined by the following table.

Using the trigger controls.   To review what you've learned about the trigger circuit and its controls (shown schematically in Fig. A.14), first set your controls as follows.

VERTICAL MODE: CH 1

CH 1 VOLTS/DIV: 0.2 (10X PROBE)

Variable (CAL): In detent and pushed in
Coupling: DC

HORIZONTAL MODE: X1

SEC/DIV: 0.5 ms

SEC/DIV CAL: In detent

TRIGGER MODE: P-P AUTO

SOURCE: CH 1
COUPLING: DC

Connect the probe to the CH 1 input connector and turn on your scope. Apply the probe tip to the PROBE ADJUST terminal and follow the directions in Exercise 5. Use the figure at the back of this appendix to remind yourself of the correct control locations.

Trigger coupling	Function
AC	Blocks dc components of the trigger signal and couples only the ac components.
LF REJ	Filters the low frequencies from the trigger signal and passes only the high frequencies (above 30 kHz). Signal is AC coupled.
HF REJ	Filters the high frequencies from the trigger signal and passes only the low frequencies (below 30 kHz). Signal is DC coupled.
DC	Passes both ac and dc trigger-signal components to the trigger circuit.

Exercise 5. TRIGGER CONTROLS

1. Move the trace to the right with the HORIZONTAL COARSE POSITION control until you can see the beginning of the signal. You'll probably have to increase the intensity to see the faster vertical part of the waveform. Watch the signal while you operate the SLOPE control. If you select the rising edge (   ), the signal on the screen starts with a positive-going slope, or rising edge; the other position (   ) makes the scope trigger on a negative-going slope, or falling edge.

2. Now turn the LEVEL control back and forth as far as it will go; you'll see the leading edge climb up and down the signal. The scope remains triggered because you are using the P-P AUTO setting. Now set input coupling to GND and notice that a bright baseline trace is present. The baseline trace appears because of the automatic-mode timer described earlier in this appendix.

3. Change TRIGGER MODE to NORM and channel 1 input coupling to DC. Now when you use the LEVEL control to move the trigger point, you'll find places where the scope is untriggered (a blank crt screen). Retrigger on the signal, then switch input coupling to GND. Notice that in NORM mode the screen is blank (no bright baseline). This is an illustration of the essential difference between normal and automatic triggering.

4. Switch input coupling back to DC and adjust the TRIGGER LEVEL control for a stable display. Set TRIGGER MODE to SGL SWP, then return channel 1 input coupling to GND. Now momentarily push in and release the single-sweep RESET button and observe that the READY-TRIG'D indicator is illumined. At this point, rotate the INTENSITY control fully clockwise to facilitate viewing the single-sweep display. Finally, move the input coupling switch from GND to DC and observe that only one sweep occurs. Press RESET each time you want to see the single-sweep display again.

5. Without a trigger signal applied to the EXT INPUT connector, it's impossible to show you the use of this trigger source, but the TRIGGER MODE, SLOPE, and LEVEL controls all work the same for either internal or external triggers. One difference between internal and external trigger sources, however, is the sensitivity of the trigger circuit. All external sources are specified and measured in voltage (say, 150 mV) while internal sources are rated in divisions. In other words, the displayed amplitude for internal signals makes a difference.

To demonstrate this, change both the VERTICAL MODE switch and the TRIGGER SOURCE switch bank to CH 1. Then set the CH 1 VOLTS/DIV switch to 0.5 V (10X PROBE) and move the TRIGGER MODE switch to NORM. Rotate the LEVEL control and notice the control range. Now change the CH 1 VOLTS/DIV switch to 0.1 V (10X PROBE) and rotate the LEVEL control. Notice that you have a broader control range now.

6. Change the SEC/DIV switch setting to 1 ms and observe the rate at which the display is updated by the recurring sweep. Now rotate the TRIGGER HOLDOFF control fully clockwise from MIN to maximum sweep-holdoff time. Notice that the display update rate is much less now. This is due to the increased holdoff time. As previously described in this appendix, variable holdoff is designed for triggering on complex waveform periods.

7. There are several other features, but they are not easily demonstrated with the PROBE ADJUST signal. You'll find them useful the first time your own measurement applications require them—applications such as video service, where the TV FIELD and TV LINE triggering modes would be needed. Also, using VERT MODE TRIGGER SOURCE when you've selected BOTH-ALT VERTICAL MODE lets you make timing measurements on two asynchronous (time unrelated) signals. Using the TRIGGER SOURCE called LINE is important for application where the signal applied to the scope's vertical input is time related to the power source, or mains, such as a power supply. Finally, keep in mind that the use of LF REJ and HF REJ TRIGGER COUPLING is ideal in noisy signal environments.

All About Probes

Connecting measurement test points to the inputs of your oscilloscope is best done with a probe like the one illustrated in Fig. A.15. Though you could connect the scope and the circuit under test with just a wire, this simplest of connections would not let you realize the full capabilities of your scope. The connection would probably load the circuit excessively, and the wire would act as an antenna, picking up stray signals such as 60-Hz power, line noise, radio transmissions, and TV stations. These signals would be displayed on the screen along with the signal of interest.

Circuit loading. Using a probe instead of a bare wire minimizes stray signals, but *circuit loading* is still an undesirable side effect. Depending on how great the loading is, circuit-loading effect modifies the circuit environment and changes the signals in the circuit under test—either a little or a lot.

Circuit loading is resistive, capacitive, and inductive. For signal fre-

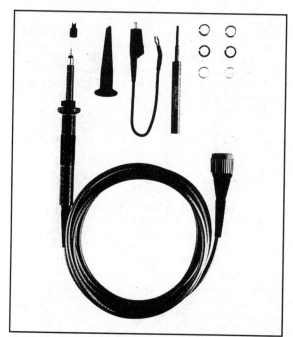

Figure A.15   Probes connect the scope and the circuit under test. Tektronix probes consist of a patented resistive cable and a grounded shield. Two P6103 10X passive probes and the accessories pictured above are supplied with every Tektronix 2225 scope. These modular, high-impedance, minimum-loading probes incorporate ruggedized probe tips and eliminate the bulky compensation box at the front of the scope. The accessories for each probe (from left to right) are: an IC tester tip cover, a retractable-hook tip, a ground lead, a compensation adjustment tool, and marker bands.

quencies under 5 kHz, the most important component of loading is resistance. In these situations, you can avoid significant circuit loading by using a probe with a resistance at least two orders of magnitude greater than the circuit impedance (100 MΩ probes for 1 MΩ sources; 1 MΩ probes for 10 kΩ sources; and so).

When you are making measurements on a circuit that contains high frequency signals, inductance and capacitance become important. You can't avoid adding capacitance when making a connection, but you can avoid adding more capacitance than necessary.

One way to do that is to use an attenuator probe. Its design greatly reduces circuit loading. Instead of loading the test circuit with capacitance from the probe tip, its cable, and the scope's input circuitry, the 10X attenuator probe introduces about ten times less capacitance—as little as 10 to 14 picofarads (abbreviated *pF* and commonly pronounced *puff*). The penalty is a reduction in test signal amplitude as seen on the screen; this is caused by the 10:1 attenuation.

The Tektronix 2225 Oscilloscope includes two P6103 10X modular construction probes that incorporate ruggedized probe tips. These probes are adjustable so you can compensate them for variations in oscilloscope input capacitance. Because the compensation adjustments are built into the probe body itself, you no longer are bothered by the bulky compensation box on the cable end that attaches to the front of the scope. *Probe compensation* is the first step in Exercise 3. Your scope has a reference signal available at the PROBE ADJUST terminal on the front panel.

Remember that when you are measuring high frequencies, the impedance (resistance and reactance) of the probe changes with frequency. The probe's specification sheet or manual will contain a chart, like the one in Fig. A.16, that shows this change. Also, when making high-frequency measurements, be sure to ground the probe securely with as short a ground lead as possible. In some very high frequency applications, a special probe socket is available that can be installed directly into the circuit. When the probe tip is inserted into the socket, the ground collar at the probe tip becomes the ground connection, and no separate lead is necessary. This type of connection gives you the shortest possible ground path, thereby minimizing impedance loading.

Measurement system bandwidth. Then there is one more probe characteristic to consider—bandwidth. Like scopes, probes have bandwidth

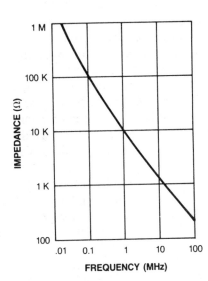

Figure A.16 Probe impedance is related to frequency, as shown in this graph. The curves plot both impedance (Z) in ohms and phase (θ) in degrees against frequency in megahertz. The plot shown is for the Tektronix P6103 probe with a 2-meter cable.

limitations; each has a specified range within which it does not attenuate the signal's amplitude more than -3 dB (0.707 of the original
value). But don't assume that a 50-MHz probe and a 50-MHz scope automatically give you 50-MHz measurement capability. The rise time
of the probe and scope combination will equal approximately the
square root of the sum of the squares of the individual rise times. For
example, if both the probe and the scope have rise times of 7.0 ns:

$$t_{r(\text{sys})} \sqrt{t^2_{r(s)} + t^2_{r(p)}}$$

$$t_{r(\text{sys})} \sqrt{49 + 49}$$

That works out to 9.9 ns, which is equivalent to a system bandwidth
of 35.36 MHz, because:

$$\text{BW(in MHz)} = 350/t_r(\text{in ns})$$

To get the full bandwidth from your scope, you need more bandwidth from the probe. It's important, therefore, to use the particular
probe designed for that instrument. For example, in the case of the
2225 Oscilloscope and the P6103 10X Passive Probe, the probe and
scope have been designed to function together, providing you a full 50-
MHz bandwidth at the probe tip.

Probe types.   Probes are classified generally as either voltage-sensing
or current-sensing. Voltage-sensing probes are divided further into
passive and active types. Refer to the following table for selecting the
probe type that will meet your measurement requirements.

Selecting a probe.   For most applications, use the probes that were
supplied with your scope. Usually these are attenuator probes. Compensation should be adjusted to ensure that the probe can faithfully
reproduce the signal for your scope. If you're not going to use the
probes that came with your scope, select a probe based on the voltage
you intend to measure. For example, if you're going to be looking at a
50-V signal and the largest vertical sensitivity is 5 V per division,
that signal will take up 10 major divisions of the screen. You would
need the attenuation of a 10X probe to reduce the amplitude of the
signal to reasonable proportions.

Proper termination to avoid unwanted signal reflections in the cable is
important. Probe and cable combinations designed to drive 1 MΩ inputs
are engineered to suppress these reflections. But, for 50-MΩ scopes, use
50-Ω probes. The proper termination also is necessary when you use a
coaxial cable instead of a probe. If you use a 50-Ω cable and a 1-MΩ
scope, be sure you also use a 50-Ω terminator at the scope input.

The probe's ruggedness and flexibility, as well as the length of its cable, can also be important. The longer the cable, the greater the capacitance at the probe tip. Remember? Check the specifications to see whether the probe bandwidth is sufficient and make sure you have the adapter and tips you'll need. Most modern probes feature interchangeable tips and adapters for many applications. Retractable hook tips let you attach the probe to most circuit components. Other adapters either connect probe leads to coaxial connectors or slip over square pins. Allegator clips for contacting large-diameter test points are another possibility.

For the reasons already mentioned (probe bandwidth, circuit loading, and termination), the best way to ensure that your scope and probe measurement system has the least distorting effect on your measurements is to use the probe recommended for your scope. And always make sure it's compensated.

Probe types	Characteristics
1X Passive, Voltage-sensing	No signal reduction, which allows maximum Voltage-sensing sensitivity at the probe tip; bandwidth typically ranges from 4 MHz to 34 MHz; high capacitance, typically 32 pF to 112 pF; signal handling to 500 V.
10X/100X/1000X Passive, Voltage-sensing, Attenuator	Attenuates signals; bandwidths to 350 MHz; adjustable capacitance; signal handling to 500 V (10X), 1.5 kV (100X), or 20 kV (1000X).
Active, Voltage-sensing, FET	Switchable attenuation; capacitance as low as 1.5 pF; more expensive, less rugged than other types; limited dynamic range; bandwidths to 900 MHz; minimum circuit loading.
Current-sensing	Measure currents from 1 mA to 1000 A; DC to 50 MHz; very low circuit loading.
High voltage	Signal handling to 40 kV.

Part II. Making Measurements

A *wave* is a disturbance traveling through a medium, and a *waveform* is a graphic representation of a wave. Like a wave, a waveform depends both on movement and on time. The ripple on the surface of a pond is a movement of water in time. The waveform on the oscilloscope screen is a movement of an electron beam over time.

Changes over time form the wave shape, the most readily identifiable characteristic of a waveform. Figure A.17 illustrates some common wave shapes.

Waveshapes tell you a great deal about the signal. Any time you see a change in the vertical dimension of a signal, you know that this amplitude change represents a change in voltage. Any time there's a flat

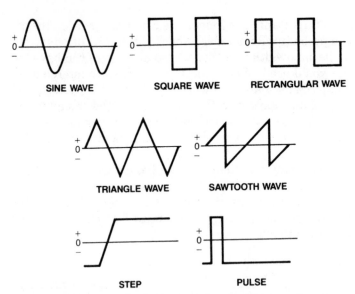

Figure A.17 Basic wave shapes include *sine* waves and various nonsinusoidal waves such as *square* waves, *rectangle* waves, *triangle* waves, and *sawtooth* waves. In a square wave, equal amounts of time are spent in each of two states (high and low). Triangle and sawtooth waves are usually the result of circuits designed to control voltage with respect to time, such as the sweep of an oscilloscope or some television circuits. In these waveforms, one or both transitions from state to state are made with a steady variation at a constant rate—a *ramp*. Changes from one state to another on all waveforms except sine waves are called transitions. The last two drawings represent aperiodic single shot waveforms. The first is a *pulse;* all pulses are marked by a rise, a time duration, and a decay. The second is a *step*—actually a single transition.

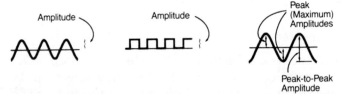

Figure A.18 Amplitude, a characteristic of all waveforms, is the amount of displacement from equilibrium at a particular point in time. Note that without a modifier, the word means the maximum change from a reference without regard to the direction of the change. In the top two drawings above (sine wave and square wave), the amplitudes are equal, even though the sine wave is larger from peak to peak. In the third drawing, an alternating current waveform is shown with its peak (or maximum) amplitude and peak-to-peak amplitude parameters annotated. In oscilloscope measurements, amplitude usually means peak-to-peak amplitude.

horizontal line, there is no change for that length of time. Straight diagonal lines mean a linear change—rise or fall of voltage at a steady rate over time. Sharp angles on a waveform mean sudden change.

But wave shapes alone are not the whole story. To completely describe a waveform, you'll want to find its particular parameters. Depending on the signal, these parameters might be amplitude, period, frequency, width, rise time, or phase. You can review these signal parameters in Figs. A.18 through A.23.

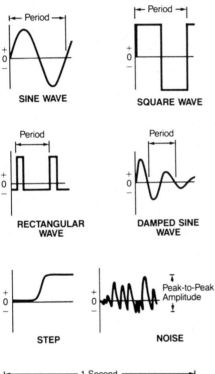

Figure A.19 Period is the time required for one cycle of a repetitive signal and is expressed in units of time. Period (its symbol is T) is a parameter whether the signal is symmetrically shaped (like the sine and square waves) or whether the signal has a more complex and asymmetrical shape (like the rectangular wave and the damped sine wave). One-time signals such as the step and uncorrelated (having no time relation) signals such as noise have no period.

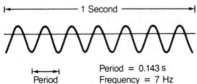

Figure A.20 If a signal is periodic it has frequency. Frequency (f) is the number of times a signal repeats itself in a second and is measured in hertz: 1 Hz = 1 cycle per second, 1 kHz (kilohertz) = 1000 cycles per second, and 1 MHz (megahertz) = 1,000,000 cycles per second. Period and frequency are reciprocal: 1/period = frequency, and 1/frequency = period. For example, a 7-Hz signal has a period of 0.143 seconds, since: 1 cycle ÷ 7Hz = 0.143s and 1 cycle ÷ 0.143 s = 7 Hz.

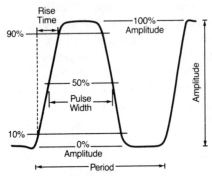

Figure A.21   The parameters of a pulse can be important in a number of different applications, including digital circuitry, X ray equipment, and data communications. Pulse specifications include transition times measured on the leading edge of a positive-going transition; this is the *rise time*. *Fall time* is the transition time on a negative-going trailing edge. *Pulse width* is measured at the 50 percent amplitude points, and *amplitude* is measured from 0 percent to 100 percent. Any displacement of the base of the pulse from zero volts is the *dc offset*.

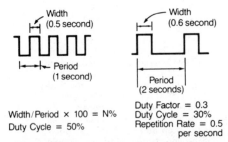

Figure A.22   Duty cycle, duty factor, and repetition rate are parameters of rectangular waves. They are particularly important in digital circuitry. *Duty cycle* is the ratio of pulse width to signal period, expressed as a percentage. For square waves, it's always 50 percent, as you can see in the top drawing. For the rectangular waveform in the bottom drawing, it's 30 percent. *Duty factor* is the same thing as duty cycle, except that it is expressed as a decimal instead of a percentage. *Repetition rate* describes how often a pulse train occurs and is used instead of frequency to describe rectangular waveforms.

Safety

Before you make any oscilloscope measurement, remember that you must be careful when you work with electrical equipment. Observe all safety precautions described in the operator's and service manuals for the equipment you're working on.

Some general rules about servicing electrical equipment are worth

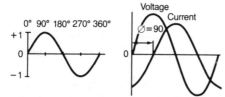

Figure A.23 Phase is best explained with a sine wave. Remember that this waveform type is based on the sine of all angles from 0° through 360°. The result is a plot that changes from 0 at 0°, 1 at 90°, 0 again at 180°, −1 at 270°, and finally 0 again at 360°. Consequently, it is useful to refer to the *phase angle* (or simply *phase*) of a sine wave when you want to describe how much of the period has elapsed. In another usage, *phase shift* describes a relationship between two signals. Picture two clocks with their second hands sweeping the dial every 60 seconds. If the second hands reach twelve at the same time, the clocks are *in phase;* if they don't, the clocks are *out of phase*. *Phase shift* expresses the amount that two signals are out of phase. To illustrate, the waveform labeled current in the drawing is said to be 90° out of phase with the voltage waveform. Other ways of reporting the same information are: the Current waveform lags the Voltage waveform by 90°, or the Current waveform has a 90° phase angle with respect to the Voltage waveform. Note that there is always a reference to another waveform; in this case, the Current waveform relative to the Voltage waveform for an inductor.

repeating here. Always service electrical devices with someone else present—don't service alone. Know the symbols for dangerous circuits and observe the safety instructions for the equipment you're working on. Don't operate or service an electrical device in an explosive atmosphere. Ground both your scope and the circuit under test to the same ground if possible. Remember that if you lose the scope's power-cord ground, all accessible conductive parts—including knobs that appear to be insulated—can give you a shock. To avoid personal injury, don't touch exposed connections or components in the circuit under test when the power is on. And remember to consult the service manual for the equipment you are working on.

Then there are a few rules about the scope itself. To avoid a shock, plug the power cord of the scope into a properly wired receptacle before connecting your probes. Use the appropriate power cord for your scope, and then only if it is in good condition. If the power cord is cracked or broken or has any pins missing, replace it. Use the correct fuse to avoid fire hazards. Don't remove covers and panels on your scope without the proper training.

Getting Started

For accurate oscilloscope measurements, make sure your system is properly set up each time you begin using your scope.

Compensating the probe. Most measurements you make with an oscilloscope require an attenuator probe, which as you learned earlier, is any probe that reduces voltage. The most common are 10X (read: ten times) passive probes. These reduce the amplitude of the signal and the circuit loading by a factor of 10:1.

Before using an attenuator probe, make sure it's compensated. Figure A.24 illustrates what can happen to the waveforms you'll see when the probe is not properly compensated.

Note that you should compensate the probe (with the accessory tip

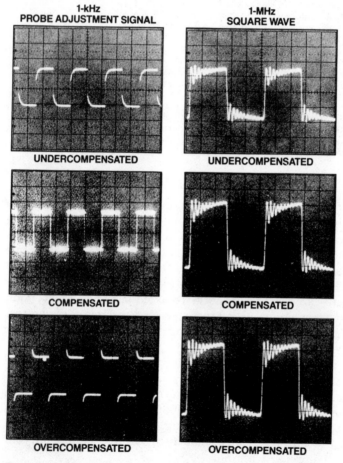

Figure A.24 Improperly compensated probes can distort the waveforms you see on the screen of your scope. In the photographs, the 1-kHz probe-adjustment signal and a 1-MHz square wave are shown as they would appear with proper and improper compensations. Notice the changes in amplitude and ringing on the 1-MHz square wave with differences in compensation.

you will use) to the vertical channel you plan to use. Don't compensate it on the one channel, then use it on another.

Checking the controls. Forgetting to compensate the probe is the most common mistake in making oscilloscope measurements. Forgetting to check the front panel controls is the second most frequent. Here are some things to check on your Tektronix 2225 scope, arranged according to functional blocks.

Check vertical system controls:

- Vertical MODE switches should be set to display the signal from the proper channel(s).

- Channel 2 MODE (center switch) should be set to NORM (unless you want it to be inverted).

- The VOLTS/DIV switches should be set to the appropriate settings. Use the VOLTS/DIV readout that matches the probe, either 1X or 10X PROBE.

- Variable CAL controls for CH 1 and CH 2 should be in their detent (calibrated) positions and pressed in (for X1 vertical magnification).

- Input coupling switches should be set for the type of signal applied to the input or for the measurements that you will make.

Check the horizontal system controls to be sure that the SEC/DIV variable (CAL) control is in its calibrated detent (fully clockwise) and that the MODE switch is set to one of the following:

- X1—When you are not making measurements requiring waveform expansion.

- ALT—When waveform points of interest are to be located and viewed on both the unmagnified and the magnified traces simultaneously.

- MAG—When you want to view only the magnified trace.

Check the trigger system controls to make sure that your scope will select the proper SLOPE on the trigger signal, that the correct operating MODE will be used, and that the appropriate SOURCE and COUPLING are selected. Also make sure that the trigger variable HOLDOFF control is set to minimum (MIN).

Handling a probe. Before you probe a circuit, make sure you have the correct probe tips and adapters for the circuits you will be working on. Tips available for the Tektronix P6103 10X probes were illustrated in Fig. A.15.

Then make sure that the ground in the circuit under test is the same as the scope ground—don't just assume that it is. The scope ground will always be earth ground as long as you're using the proper power cord and plug. Check the circuit ground by first attaching the ground lead of your probe to a known earth ground. Then touch the probe tip to the point you think is ground. Do this before you make a *hard* ground—that is, attaching the ground strap of your probe.

If you're going to be probing a lot of different points in the same circuit and measuring frequencies less than 5 MHz, you can ground that circuit to your scope once instead of each time you move the probe. To do this, just connect the circuit ground to the jack marked on the front panel.

Measurement Techniques

Rather than showing how to make every possible measurement, this chapter describes common measurement techniques you can use in many applications.

The foundations: amplitude and time measurements.   The two most basic measurements you can make are amplitude and time. Almost every other measurement is based on one of these two fundamental techniques.

Since the oscilloscope is a voltage-measuring device, voltage is shown as amplitude on your scope screen. Of course, voltage, current, resistance, and power (watts) are related in the following way:

$$\text{Current} = \frac{\text{Voltage}}{\text{Resistance}} \qquad \text{Current} = \frac{\text{Power}}{\text{Voltage}}$$

$$\text{Resistance} = \frac{\text{Voltage}}{\text{Current}} \qquad \text{Voltage} = \frac{\text{Power}}{\text{Current}}$$

$$\text{Voltage} = \text{Current} \times \text{Resistance} \qquad \text{Power} = \text{Current} \times \text{Voltage}$$

Amplitude measurements are best made with a signal that covers most of the screen vertically. This is done because the more screen area you use, the better measurement resolution you can achieve. Use Exercise 6 for practicing amplitude measurements.

Time measurements are also more accurate when the signal covers a large area of the screen. Continue with the setup you had for the amplitude measurement (Exercise 6), but now use Exercise 7 to make a period measurement.

Exercise 6. AMPLITUDE MEASUREMENTS

1. Connect your probe to the CH 1 connector and attach the tip to the PROBE AD-JUST terminal. Attach the probe ground lead to the collar of the CH 2 connector. Make sure your probe is compensated and all variable (CAL) controls are in their detents.

2. Set the VERTICAL MODE switch to CH 1 and the channel 1 input coupling switch to AC. The TRIGGER MODE switch should be set to NORM—for normal triggering—and the TRIGGER SOURCE switch set to CH 1.

3. Use the TRIGGER LEVEL control to obtain a stable trace and move the CH 1 VOLTS/DIV switch until the probe-adjustment square wave is about five divisions high. Now turn the SEC/DIV switch until about one cycle of the waveform is on your screen. The settings should be 0.1 V (10X PROBE) on the VOLTS/DIV switch and 0.1 ms on the SEC/DIV switch.

4. Now use the channel 1 POSITION control to move the square wave so that its bottom is aligned with a convenient horizontal graticule line that lets you approximately center the waveform vertically. Use the HORIZONTAL COARSE POSITION control to move the signal so that either the top or the bottom of one cycle intersects the center vertical graticule line.

5. You now can make the measurement. Count major and minor divisions up the center vertical graticule line and multiply by the VOLTS/DIV setting. For example, 5.4 division times 0.1 V per division equal 0.54 V. If the voltage of the probe adjustment square wave in your scope is different from this example, that's because this signal is not a critical part of your scope and tight tolerances with exact calibration are not required.

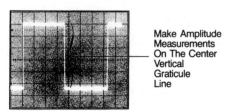

Make Amplitude
Measurements
On The Center
Vertical
Graticule
Line

Exercise 7. TIME MEASUREMENTS

Time measurements are best made with the center horizontal graticule line. In this exercise, to make a period measurement, use the existing instrument settings from Exercise 6. Then, with the HORIZONTAL COARSE and FINE POSITION controls, line up one rising edge of the square wave with the graticule line that's second from the left side of the screen. Make sure that each rising edge intersects the center horizontal graticule line.

Count major and minor divisions across the center horizontal graticule line from left to right as shown in the photo. Multiply by the SEC/Div setting; for example

(Continued)

7.2 division times 0.1 ms per division equals 0.7 percent ms. If the period of the probe adjustment square wave in your scope is different from this example, remember that this signal is not a critical part of the calibration of your scope.

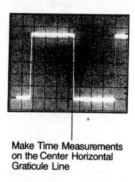

Make Time Measurements
on the Center Horizontal
Graticule Line

Frequency and other derived measurements.   The voltage and time measurements you just made are two examples of *direct measurements*. Once you've made a direct measurement, you can calculate *derived measurements*. Frequency, derived from period measurements, is one example. While period is the length of time required to complete one cycle of a periodic waveform, frequency is the number of cycles that take place in a second. The measurement unit is hertz (one cycle per

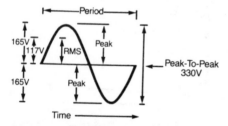

Figure A.25   Derived measurements are the result of calculations after making direct measurements. For example, alternating current measurements first require an amplitude measurement. The easiest place to start is with a peak-to-peak amplitude measurement of the voltage—in this case, 330 volts—because peak-to-peak measurements ignore positive and negative signs. The *peak voltage* is one-half that value (when there is no DC offset), and is also called *maximum value;* in this case it's 165 V. The average *value* is the total area under the voltage curves divided by the period in radians; in the case of a sine wave, the average value is zero because the positive and negative values are equal. But in some applications such as power, the average value is determined to be: $V_{avg} = 0.318 \times V_{p-p}$. The rms (root-mean-square) voltage for this sine wave—which represents the line voltage in the United States—is: $V_{rms} = (V_p/\sqrt{2} = (165/1.414) = 117$ V. You get from peak-to-peak to rms $V_{rms} = (V_{p-p}/(2\sqrt{2})$.

second) and it's the reciprocal of period. So a period of 0.001 second (or 1 ms) has a frequency of 1000 Hz (or 1 kHz).

More examples of derived measurements are the alternating-current measurements illustrated in Fig. A.25.

The following table is a convenient reference for calculating the four main waveform parameters.

Calculating Waveform Parameters

To calculate this amplitude	Multiply this known value	By this conversion factor
RMS	AVG	1.110[a]
		2.220[b]
	PEAK	0.70
	P-P	0.354
AVG	RMS	0.900[a]
		0.450[b]
	PEAK	0.637
	P-P	0.318
PEAK	RMS	1.414
	AVG	1.570
	P-P	0.500
P-P	RMS	2.828
	AVG	3.140
	PEAK	2.00

[a]For full-wave rectification
[b]For half-wave rectification

Phase measurements. You know that a waveform has phase, the amount of time that has passed since the cycle began, measured in degrees. Waveforms can also be related by phase shift, and there are two ways to measure it. The first way to measure phase shift between two waveforms is to connect one waveform to each vertical input of a dual channel scope and view them directly either in the chopped or in the alternated vertical mode. Trigger on the channel containing the reference signal. Adjust the TRIGGER LEVEL control for a stable display and measure the waveform period. Adjust the sweep speed so that you have a display similar to the second drawing in Fig. A.23. Then measure the horizontal distance between the same points on the two waveforms. The phase shift is the time difference between the points divided by the waveform period, then multiplied by 360 to give degrees.

The second way to measure phase shift is to obtain a Lissajous pattern. Notice that the CH 1 and CH 2 vertical input connectors also are labeled X and Y respectively and that the last position on the SEC/DIV switch is X-Y. This X-Y setting disables the scope's time base.

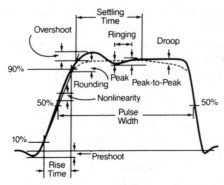

Figure A.26 Real pulse measurements include a few more parameters than those for an ideal pulse. You'll find several in the diagram above. *Preshoot* is a change of amplitude in the opposite direction that precedes the rising step. *Overshoot* and *rounding* are changes that occur after the initial transition. *Ringing* is a series of amplitude changes—usually a damped sinusoid—that follows the front corner. All are expressed as percentages of amplitude. *Setting time* expresses how long it takes the pulse to stabilize at its maximum amplitude. *Droop* is a decrease in the maximum amplitude with time. *Nonlinearity* is any variation from a straight line drawn through the 10 percent and 90 percent points of a transition.

When X-Y is selected, the channel 2 input signal is still the vertical axis of the scope's display, but now the channel 1 signal becomes the horizontal axis.

In the X-Y mode, with one sinusoidal waveform on each channel, your screen displays a Lissajous pattern (named for French physicist Jules Antoine Lissajous and pronounced *LEE-sa-zhoo*). The shape of the pattern indicates the phase shift between the two signals. Some examples of Lissajous patterns are shown in Fig. A.27.

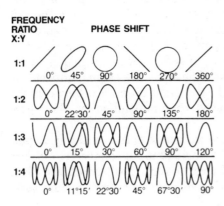

Figure A.27 Frequency measurements with Lissajous patterns require a known sine wave on one channel. If there is no phase shift, the ratio between the known frequency (usually applied to the X input) and the unknown frequency (applied to the Y input) corresponds to the ratio between the number of vertical loops and the number of horizontal loops in the pattern. When the frequencies are the same, only the shifts in phase affect the pattern. In the drawings here, both phase and frequency differences are shown.

Note that Lissajous measurements with general-purpose oscilloscopes are usually limited by the frequency response of the horizontal amplifier (typically designed with a far lower bandwidth than vertical channels). Specialized X-Y scopes or monitors have almost identical vertical and horizontal systems.

Exercise 8. DERIVED MEASUREMENTS

With the period measurement you made in Exercise 7, calculate the frequency of
the probe adjustment square wave. For example, if the period you measured is 0.72
ms, then the frequency is the reciprocal:

$$f = \frac{1}{T} = \frac{1}{0.00072 \text{ s}}$$
$$= 1389 \text{ Hz}$$
$$= 1.389 \text{ kHz}$$

Other derived measurements you can make are duty cycle, duty factor, and repetition rate. Duty cycle is the ratio of pulse width to signal period expressed as a
percentage: 0.5 ms divided by 1 ms, or 50 percent. (But you knew that; it's always 50
percent for square waves.) Duty factor, though, is 0.5. The repetition rate describes
how often a pulse train occurs. For square waves, repetition rate and frequency are
equal. Your probe adjustment signal might differ slightly from this example; if so,
calculate the duty cycle, duty factor, and repetition rate for it. You can also calculate
the peak, peak-to-peak, and average voltages of the probe-adjustment square wave
in your scope. Remember that you need both the alternating and direct components
of the signal to make these measurements, so be sure the input coupling switch on
the vertical channel you're using is set to DC.

Use the directions in Exercise 9 to make a pulse width measurement on the probe-
adjustment square wave.

Exercise 9. PULSE WIDTH MEASUREMENTS

To measure the pulse width of the probe-adjustment square wave quickly and easily, set your scope to trigger on and display channel 1. Your probe should still be
connected to the CH 1 BNC connector and the PROBE ADJUST terminal from the
previous exercises. Use a SEC/DIV setting of 0.1 ms with HORIZONTAL MODE at
X1; use P-P AUTO triggering on the positive-going (   ) slope. Adjust the TRIGGER
LEVEL control to get as much of the leading edge on screen as possible. Switch the
CH 1 input coupling to GND and center the baseline on the center horizontal
graticule line. Now switch to AC coupling, because that will center the signal on the
screen and let you make a pulse-width measurement at the 50% point of the waveform. Use the HORIZONTAL COARSE and FINE POSITION controls to line up the
50 percent point with the first graticule line from the left side of the screen. Now you
can count divisions and subdivisions across the center horizontal line until you reach
the falling edge. Then multiply that value by the SEC/DIV switch setting to find the
positive half pulse width. Switch the TRIGGER SLOPE switch to the falling edge
(   ). Repeat the same measurement process to find the negative-half pulse width
(see Fig. A.26).

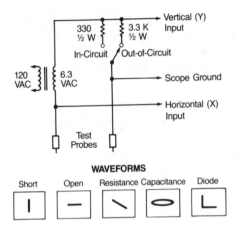

Figure A.28 X-Y component checking requires the transformer circuit shown above. With it connected to your scope and the scope in the X-Y mode, patterns like those illustrated indicate the condition of the component. The patterns shown can be seen when the components are tested out of the circuit; in-circuit component patterns differ because of resistors, capacitors, and other devices connected to the component under test.

X-Y measurements. Finding the phase shift of two sinusoidal signals with a Lissajous pattern is one example of an X-Y measurement. The X-Y capability can be used for other measurements as well. Lissajous patterns can also be used to determine the frequency of an unknown signal when you have a known signal on the other channel. This is a very accurate frequency measurement as long as the known signal is accurate and both signals are sine waves. The patterns you can see are illustrated in Fig. A.27, where the effects of both frequency and phase difference are shown.

Component checking in a service or production situation is another X-Y application and requires only a simple transformer circuit like the one shown in Fig. A.28.

You'll find many other applications for X-Y measurements in television servicing, engine analysis, two-way radio servicing, and more. Any time you have physical phenomena that are interdependent and not time-dependent, X-Y measurements are the answer. Aerodynamic lift and drag, motor speed and torque, or pressure and volume of liquids and gasses are more examples. With the proper transducer, you can use your scope to make any of these measurements.

Differential measurements. The ADD and the CH 2 INVERT VERTICAL MODE functions on your 2225 Oscilloscope allow you to make differential measurements. Often differential measurements permit eliminating undesirable components from a signal. If you have a signal that's very similar to the unwanted noise, the setup is simple. Put the signal with the spurious information on channel 1. Connect the signal that is like it, but with unwanted components, on channel 2. Set both input coupling switches to DC (use AC if the DC components of the signal(s) are too large) and select a dual channel vertical mode by moving the VERTICAL MODE switches to BOTH-ALT (or CHOP depending on signal frequency).

Now set both VOLTS/DIV switches so that the noise components of the two signals are about equal in amplitude. Then you can move the right VERTICAL MODE switch to ADD and the middle switch to CH 2 INVERT so that the common-mode signals have opposite polarities.

If you use the CH 2 VOLTS/DIV switch and its CAL control for maximum cancellation of the common signal, the trace that remains on screen will contain only the desired part of the channel 1 input signal. The two common-mode signals have cancelled each other out, leaving only the difference between the two. An example is shown in Fig. A.29.

Using the Z-axis. Remember from Part I that the crt has three axes of information: X is the horizontal axis, Y is the vertical axis, and Z is the intensity axis. The Z-axis input lets you change the brightness of the signal on the screen with an external signal. On the 2225, the Z-axis input will accept a signal of up to 400 V (dc + peak ac) through a usable frequency range of DC to 5 MHz. Positive voltages decrease brightness and negative voltages increase it; 5 V causes a noticeable change.

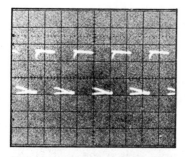
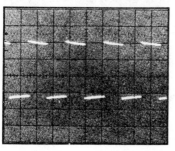

Figure A.29 Differential measurements allow you to remove unwanted information from a signal any time you have another signal that closely resembles the unwanted components. For example, the first photo shows a 1-kHz square wave contaminated by a 60-Hz sine wave. Once the common-mode component (the sine wave) is input to channel 2 and that channel is inverted, the signals can be added by selecting ADD VERTICAL MODE. The result is shown in the second photo.

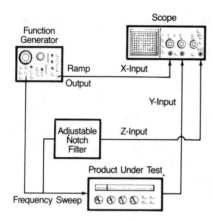

Figure A.30 Using the Z-axis can provide additional information on the scope screen. In the setup drawn above, a function generator sweeps through the frequencies of interest during product testing—20 Hz to 20,000 Hz, in this case. Then an adjustable notch filter is used to generate a marker—at 15 kHz, for instance—and this signal is applied to the Z-axis input to brighten the trace. This allows the tester to evaluate product performance at a glance.

The Z-axis input is an advantage to users who have set up their instruments for a long series of tests. One example is the testing of high-fidelity equipment illustrated by Fig. A.30.

TV Triggering. The NTSC composite-video waveform consists of two fields, each containing 262½ lines. Many scopes offer television triggering to simplify viewing video signals. Usually, however, the scope will only trigger on fields at some sweep speeds and on lines at others. The 2225 will let you trigger on either lines or fields at any sweep speed.

To look at TV fields with a 2225 Oscilloscope, use the TV FIELD TRIGGER MODE.

This mode allows the scope to trigger at the field rate of the composite video signal on either field 1 or field 2. Since the trigger system cannot recognize the difference between the two fields, it will trigger alternately on them. The display will be confusing if you look at one line at a time.

To prevent this, add more holdoff time using one of two methods. Either use the HOLDOFF control, or simply switch the VERTICAL MODE to display BOTH channels. That makes the total holdoff time for one channel greater than one field period. Then position the unused vertical channel off-screen to avoid confusion.

It is also important to select the trigger slope that corresponds to the edge of the waveform where the sync pulses are located. Picking a negative slope for triggering on TV sync signals will show as many sync pulses as possible.

When you want to observe the TV line portion of the composite video signal, use the TV LINE TRIGGER MODE and get a stable display by triggering on the horizontal synchronization pulses. It is usually best to select the blanking level of the sync waveform so that the vertical field rate will not cause double triggering.

Waveform expansion for detailed analysis. The unmagnified sweep—that's the trace you see when HORIZONTAL MODE is set to X1—is the horizontal function you'll need for most applications. Best measurement accuracy is achieved by setting the SEC/DIV control for the fastest sweep that will display the interval of interest. And remember that the variable control (CAL) should be in its detent (fully clockwise).

On the 2225, there are also two magnified modes, ALT and MAG. These modes take the measurement of time a step further. Because they expand the unmagnified trace (that is, the *normal* trace), the ALT and MAG modes give you the ability to make a variety of measurements that previously were only possible on scopes having dual time bases. These include measurements such as digital system timing, television signal analysis, and waveform comparisons.

When ALT is chosen, both the normal and the magnified waveforms appear together on the crt screen. Since you can use the 2225 Oscilloscope to display the normal and the magnified waveforms from both channel 1 and channel 2, it's possible to have four traces on the crt screen at one time.

To prevent overlapping traces in this situation, there is an additional position control—labeled TRACE SEP—which adjusts the vertical separation between the normal and the magnified traces. By using the TRACE SEP control together with the two VERTICAL POSITION controls, you can place all four traces on the screen without confusion.

When MAG is selected, only the magnified trace is displayed on the screen. This is useful for eliminating unwanted clutter from the crt when you are making accurate time measurements or are looking at waveform details.

In either ALT or MAG MODE, the amount of waveform expansion is determined by the setting of the HORIZONTAL MAG switch located beneath the SEC/DIV control. Three magnifications are selectable—5X, 10X, and 50X. Magnification increases the sweep speed by the set amount without changing the basic SEC/DIV setting. For example, if the SEC/DIV control were set to 50 μs, with ALT MODE chosen and 10X MAG selected, the sweep speed of the magnified trace would be 5 μs per division. Having the ability to select various combinations of waveform expansion and SEC/DIV control settings lets you extend the timebase range out to a maximum of 5 ns per division.

The registration marker that links the timing of the normal and the magnified traces with each other is the center vertical graticule line. The intersections of that line with the normal and the magnified waveforms represent the same point in time (from sweep start) on both traces. With the center vertical graticule as the reference line, you have the ability to investigate waveform details around any point you choose on the normal trace. And because it's expanded, the magnified waveform makes your time measurement easier to perform as well as enhances the precision.

Using the horizontal magnification modes. Often the magnification modes ALT and MAG can be used to enhance the precision of a particular measurement such as rise time—especially for faster pulses. The measurement in Exercise 10 is a good example. Making an accurate assessment of rise time on these pulses would be a more difficult task if the alternate magnification feature were not available. This is because the transition edge of interest could not be expanded enough along the time axis so that it occupies two or more divisions.

Using the horizontal magnification modes on the Tektronix 2225 Oscilloscope for making time measurements is a simple technique that is easily learned. Generally, the steps are: (1) display the waveform in X1 HORIZONTAL MODE at the fastest sweep speed that shows the area you wish to inspect, (2) use the COARSE POSITION control to move the area of interest to the center vertical graticule line, (3) switch HORIZONTAL MODE to ALT, and (4) set the MAG switch to the highest magnification that completely shows the area of interest on the screen.

At this point you can either leave the MODE switch at ALT or set it to MAG, depending on the particular measurement to be made and the amount of distracting clutter you wish to remove from the screen. Switching to MAG causes the normal (unmagnified) trace to disappear. Then align the magnified trace to the desired point (using the FINE POSITION control) and make your time measurement on the magnified waveform.

Expanding the trigger point is another example that demonstrates the usefulness of the horizontal magnification modes. Just set the beginning of the normal trace to the center vertical graticule line and select ALT.

Then expand the magnified trace an appropriate amount (X5, X10, or X50). With this technique, you can view the trigger point in greater detail and adjust the trigger level to exactly where you want it.

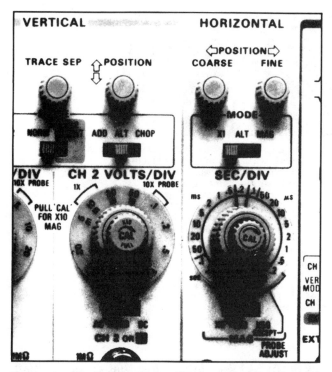

Figure A.31 The horizontal magnification controls on the 2225 Oscilloscope are shown in this photograph. They include: COARSE and FINE POSITION (top right); MODE (X1-ALT-MAG); SEC/DIV; and MAG (X5-X10-X50). One other control associated with the horizontal magnification function is TRACE SEP (top left). It moves the magnified trace vertically with respect to the unmagnified trace.

Figure A.31 shows the front-panel controls associated with the horizontal magnification function, and Fig. A.32 illustrates two measurement applications.

High sensitivity and vertical expansion. Besides horizontal magnification, there's also a vertical magnification function. It is initiated by pulling the VOLTS/DIV CAL (X10 PULL) knob out towards you. Vertical magnification expands the waveform amplitude and increases the sensitivity of the active setting on the VOLTS/DIV switch by a factor of 10. For the 5 mV per division setting on the 2225, vertical magnification increases sensitivity to a maximum of 500 μV per division. This function is especially helpful for triggering on, displaying, and making measurements on low-level signals.

Typical applications in which vertical magnification can improve voltage measurement resolution involve those that employ transduc-

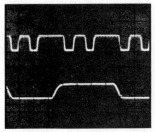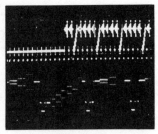

Figure A.32 Horizontal alternate magnification measurements are fast and simple to make. One use, examining timing in a digital circuit, is demonstrated in the left photograph. Suppose you need to check the width of one pulse in a pulse train like the one shown. To make sure you are measuring the correct pulse, you must look at a large portion of the signal. But to measure one pulse accurately, you need a faster sweep speed. Looking at the big picture simultaneously with an enlargement of a small part of the signal is a simple task with the horizontal alternate magnification feature. A second example is shown in the right photograph. Here, triggering is on one field of a composite video signal—displayed by the top (unmagnified) waveform. The bottom (magnified) waveform was attained by setting ALT HORIZONTAL MODE and X50 MAG. With the COARSE and FINE POSITION controls you can walk through the field and look at each line individually.

ers and similar sensors. Transducers are devices that generate small electrical signals proportional to the physical phenomena being observed. For example, fluctuations detected by a pressure transducer are converted to voltages that correspond to the pressure levels. These low-level signals then are applied to the input connectors of the oscilloscope and displayed on the crt.

Other applications in which vertical magnification is useful include the measurement of power-supply ripple. Ripple is the portion of a supply's output voltage that is harmonically related in frequency to the input power and to any internally generated switching frequency. And vertical magnification is also an effective function for measuring noise and common-mode rejection ratio.

When X10 vertical magnification is activated on a particular input channel, bandwidth on that channel is reduced to 5 MHz. Bandwidth limiting is effective in eliminating or reducing unwanted high-frequency noise components that may be present on an input signal. Since this function can be independently selected on channel 1 and channel 2, you therefore have the ability to limit the bandwidth on one channel without affecting the bandwidth on the other channel. Independent bandwidth limiting of each input channel gives you greater versatility when you are comparing two different signals.

There are other advantages in using the X10 vertical magnification function to limit bandwidth. It lowers the trigger bandwidth, which reduces unwanted noise triggering. Thus it gives you more stable triggering on low-level signals—especially those less than 5 mV p-p. Using X10 vertical magnification also lets you continue using a high-input-resistance 10X probe for measuring low-level signals instead of a 1X probe. Because the 10X probe has much less loading effect on the circuit under test than a 1X probe, the displayed signal is a more faithful reproduction of the actual input signal, and any measurements you make will naturally be closer to what is really happening in the circuit under test.

Practice measurements.   Having the ability to view a signal at two different sweep speeds makes time measurement easier. The normal trace shows you a large slice of time on the signal being examined. And the magnified trace—the one with the faster sweep speed—expands the normal waveform to allow inspection of any portion in greater detail. You'll find this capability useful in many measurement applications.

This discussion should have started you thinking about other uses for the magnified modes on the 2225 Oscilloscope. Now it's time to practice. As you use the scope for Exercise 10, you'll find that the procedure is very easy to perform.

Note: Alt. Magnification on the 2225 places the magnified trace above the nonmagnified trace, which is opposite from what is shown in Exercise 10.

Exercise 10. MEASUREMENTS USING HORIZONTAL ALTERNATE MAGNIFICATION

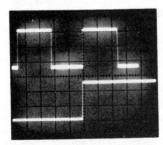 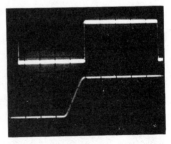

Rise Time

1. Connect your probe to the CH 1 connector and attach the probe tip to the PROBE ADJUST terminal. Hook the ground strap onto the collar of the CH 2 BNC connector and make sure the probe is compensated for channel 1.

(Continued)

2. Use these control settings: VERTICAL MODE to CH 1; CH 1 VOLTS/DIV to 0.2 (10X PROBE); CH 1 variable in CAL detent and pushed in (no vertical magnification); CH 1 input coupling to AC; HORIZONTAL MODE to X1; SEC/DIV to 0.1 ms and its variable control in CAL detent; MAG to X5; TRIGGER SLOPE to the falling edge ( ); MODE to P-P AUTO; SOURCE to either CH 1 or VERT MODE; and COUPLING to AC.

3. If necessary, adjust the TRIGGER LEVEL control for a stable display, then position the waveform in the upper half of the screen. Switch the HORIZONTAL MODE to ALT and use the TRACE SEP control to move the magnified sweep to the lower half of the screen.

4. With the COARSE and FINE POSITION controls, place the rising edge of a pulse on the upper (unmagnified) trace and the rising edge of the corresponding pulse on the lower (X5 magnified) trace along (or nearly along) the center vertical graticule line. Recall that the center vertical graticule line is the registration mark representing the same point in time (from the start of the sweep) along both the unmagnified and the magnified traces. Now your screen should look like the first photo.

5. Increase the sweep speed a step at a time by rotating the SEC/DIV switch clockwise; stop when there is one complete rising edge remaining on the upper trace. For this particular signal, the setting should be 50 μs. Again, align the rising edges of both traces to the center vertical graticule line. Then, while watching the lower trace, switch MAG to X10 and to X50; notice what happens to the rising edge in the magnified trace. Your screen should look like the second photo.

6. Eliminate the upper trace by switching MODE to MAG. Use the TRACE SEP control to center the waveform vertically. Set the CH 1 VOLTS/DIV switch to 50 mV (10X PROBE), then rotate its variable control counterclockwise out of detent. Use that control alternately with the TRACE SEP control to obtain an exact 5-division display with the zero reference aligned to the 0 percent graticule line and the maximum value aligned to the 100 percent graticule line.

7. Adjust the FOCUS and INTENSITY controls for a sharp, bright trace and then position the waveform so that it intersects a vertical graticule line at the 10 percent marking. Now your screen should look like the third photo.

8. Count the horizontal divisions between the points where the waveform crosses the 10 percent and 90 percent markings. In this example, it's 1.7 divisions. To calculate the rise time, multiply this measured distance by the SEC/DIV switch (50 μs per division), then divide the product by the magnification factor (50). Your answer should be a rise time of 1.7 μs.

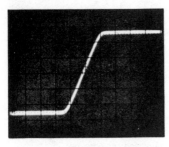 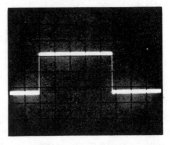

9. One last word on rise-time measurements. The accuracy of the measurement you make depends on both the human visual error and the performance of your scope. In the next section, you'll find a description of how the scope's own rise time affects measurements results.

Pulse Width

1. Perform steps 1 through 3 in the preceding rise-time measurement exercise. Switch HORIZONTAL MAG to X10.

2. Decrease the sweep speed a step at a time by rotating the SEC/DIV switch counterclockwise; stop when there is at least one complete pulse (one half of a cycle) on the lower trace. For this particular signal, the setting should be either 0.5 ms or 1 ms, depending on the exact signal frequency in the oscilloscope you're using. Eliminate the upper trace by switching MODE to MAG.

3. Switch the channel 1 input coupling to GND. Use the TRACE SEP control to align the baseline trace exactly on the center horizontal graticule line, and then switch input coupling back to AC. This centers the waveform vertically.

4. With the COARSE and FINE POSITION controls, horizontally move the pulse to be measured so that its left-most edge intersects a convenient calibration mark on the center horizontal graticule line (see photo). This step properly positions the pulse on the screen so that the pulse width can be measured at the 50 percent amplitude level.

5. Count the horizontal divisions (along the center horizontal graticule line) between the two pulse edges. In this example, it's 4.8 divisions. To calculate the pulse width, multiply the measured distance by the SEC/DIV switch setting (1 ms per division), then divide the product by the magnification factor (10). Your answer should be a pulse width of 0.48 ms.

Scope Performance

There are two aspects to oscilloscope performance: the design parameters of the instrument and its conformance to those parameters at the time you are making measurements. Making the instrument conform to its design parameters simply means calibration—including making sure the probe is properly compensated as you've done many time already. But even with proper calibration, the designed performance will have some effect on your measurements.

Square-wave response and high-frequency response. The design of amplifiers, like those in a scope's vertical channels, involves some compromise between the circuit's high frequency response and its handling of signals with square transitions. The frequency response can be extended with high frequency compensation. But too much compensation results in overshoot on a step, and too little extends the measured rise time. The best rise times without overshoot are achieved when the high-frequency response its critically damped, causing the

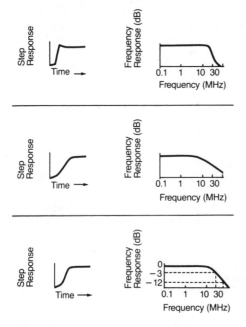

Figure A.33   High-frequency compensation in a vertical amplifier affects the rise time of square waves measured by the scope. With too much high-frequency compensation, the rise times will show overshoot and possible ringing, as in the top drawing. With too little, as shown in the second drawing, the rise times roll off the edges of the square wave. A critically damped frequency response is best, as in the third drawing.

frequency response to fall off smoothly. Figure A.33 illustrates the effects of high-frequency compensation.

Instrument rise time and measured rise time.   The rise time of an oscilloscope is a very important specification, because its rise time affects the accuracy of the rise times it measures, as expressed by this approximation:

$$t_{r(\text{measurement})} = \sqrt{t_{r(\text{signal})}^2 + t_{r(\text{system})}^2}$$

In practical terms this means that the accuracy of a measured signal is predictable and depends on how much faster your scope is than the rise times you're measuring. If the scope is five times faster than the signal, the measurement error can be as low as 2 percent. For measurement accuracies of 1 percent, you will need a scope seven times faster than the signal it measures, as shown in the chart in Fig. A.34.

Bandwidth and rise time.   The vertical channels of an oscilloscope are designed for a broad bandpass, generally from some low frequency (dc) to a much higher frequency. This is the oscilloscope's bandwidth, spec-

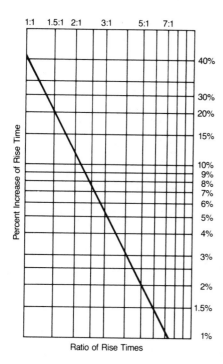

Figure A.34 Measured rise-time errors depend on the ratio of the scope's rise time to the actual rise time of the signal being measured. As you can see from the graph, when the scope is five times faster, the error is a 2 percent increase in measured rise time. If the rise times are equal, the error increases to 41 percent.

ified by listing the frequency at which a sinusoidal input signal is attenuated to 0.707 of the lower reference frequencies—the −3 dB point. For older instruments, specifications cited both a low and a high −3 dB point. Modern instruments, however, have a relatively flat frequency response down to 0 Hz (dc), so only the upper number is quoted as the bandwidth.

A bandwidth specification gives you an idea of the instrument's ability to handle high-frequency signals within a specified attenuation. But bandwidth specifications are derived from the instrument's ability to display sine waves. A 35 MHz scope will show a 35 MHz sine wave with only −3 dB attenuation, but the effects on a square wave at or near the scope's upper bandwidth limit will be much more severe, because high-frequency information in the square wave cannot be accurately reproduced by the scope. See Fig. A.35 for an example.

The frequency response of most scopes is designed with a constant that lets you relate the bandwidth and rise time of the instrument. This constant is 0.35, and the rise time and bandwidth are related by this approximation:

$$t_r = \frac{0.35}{\text{BW}}$$

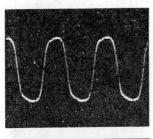

15-MHz
Square wave on a
35 MHz
Oscilloscope

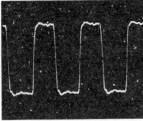

15-MHz
Square wave on a
50 MHz
Oscilloscope

Figure A.35 Bandwidth specifications are based on the scope's ability to reproduce sine waves. The upper bandwidth is the frequency at which a sine wave is reduced to 0.707 of the amplitude shown at lower reference frequencies. Though this specification tells you how well the instrument reproduces sine waves, not every signal you examine is sinusoidal. Square waves, for example, have a great deal of high-frequency information in their rising and falling edges that will be lost as you approach the bandwidth limits of the instrument. To illustrate, the two crt photos show a 15-MHz square wave reproduced by a 35-MHz oscilloscope (top) and a 50-MHz oscilloscope (bottom).

A simple way to apply the formula is:

$$t_r(\text{in ns}) = \frac{350}{\text{BW(in MHz)}}$$

For the Tektronix 2225 Oscilloscope with a bandwidth of 50 MHz, the measurement system rise time is 7.0 ns.

Conclusion

This conclusions your introduction to oscilloscopes and the measurements you can make with scopes. You've done well to progress this far, but this appendix can only introduce the concepts and measurement techniques. With practice and experience, you'll find yourself making faster and more accurate measurements. Then you, too, will find that using an oscilloscope becomes second nature. For more information about oscilloscope basics, request a copy of the 2225 Technique Brief series.

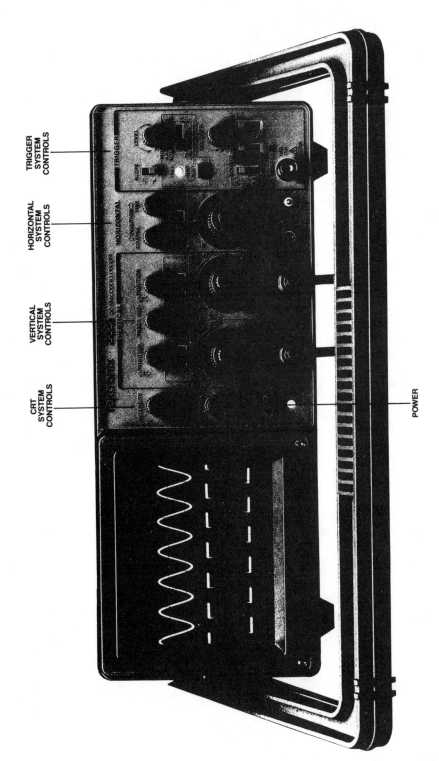

TRIGGER
SYSTEM
CONTROLS

HORIZONTAL
SYSTEM
CONTROLS

VERTICAL
SYSTEM
CONTROLS

CRT
SYSTEM
CONTROLS

POWER

293

B

Impedance Matching and Signal Level Adjustment

Networks made up of resistances in various configurations are used to match impedances between an instrument and the device tested. Such networks may be used to control precisely the signal level to the measuring instrument. These networks are usually constructed in the L, T, H, or π configuration. The resistances may be mounted simply on a circuit board with connectors on each end or they may be placed in a shielded metal box or cylinder with connectors on each end. A coaxial 75 ohm version is shown in Fig. B.1 for a T type pad with attenuation vs resistance values given in Table B.1. Figure B.2 and Table B.2 are for the 50 ohm attenuator.

The values for a 75 ohm to 50 ohm impedance matching L pad are given in Fig. B.3. This impedance-matching pad has a thorough loss of 5.7 dB. This type of pad has to have each end terminated in the appropriate impedance which may be an instrument's or tested device terminal impedance.

For audio applications a balanced or symmetrical pad is often more appropriate. The T type pad can be re-configured to an H type by taking ½ the T pad R_1 value, making it the H pad R_1 value shown in Fig. B.4. Also, the impedance values for a 75 ohm T pad can be scaled up to the 600 ohm standard audio impedance by calculating the scale factor

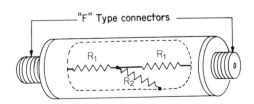

"F" Type connectors

Figure B.1

TABLE B.1

dB	R_1	R_2
1	4.3	650
2	8.6	323.0
3	12.8	213.0
4	17.0	157.3
5	21.0	123.4
6	25.0	100.5
7	28.7	83.7
8	32.2	71.0
9	35.7	61.0
10	39.0	52.7
11	42.0	46.0
12	45.0	40.2
13	47.6	35.4
14	50.0	31.2
15	52.3	27.5
16	54.5	24.4
17	56.4	21.5
18	58.2	19.2
19	60.0	17.0
20	62.5	15.2

TABLE B.2

dB	R_1	R_2
1	2.9	433.3
2	5.7	215.3
3	8.5	142.0
4	11.3	104.9
5	14.0	82.3
6	16.7	67.0
7	19.1	55.8
8	21.5	47.3
9	23.8	40.7
10	26.0	35.1
11	28.0	30.7
12	30.0	26.8
13	31.7	23.6
14	33.3	20.8
15	34.9	18.3
16	36.3	16.3
17	37.6	14.3
18	38.8	12.8
19	40.0	11.3
20	41.7	10.1

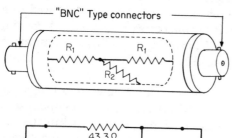

"BNC" Type connectors

Figure B.2

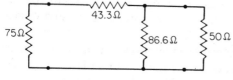

43.3 Ω   75 Ω   86.6 Ω   50 Ω

Figure B.3

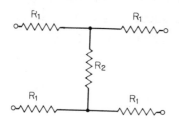

R_1 R_1 R_2 R_1 R_1

Figure B.4

TABLE B.3

dB	R_1	R_2
1	17.2	5200.0
2	34.4	2584
3	51.2	1704.0
4	68.0	1258.4
5	84.0	987.2
6	100.0	804.0
7	114.8	669.6
8	128.8	568.0
9	142.8	488.0
10	156.0	421.6
11	168.0	368.0
12	180.0	321.6
13	190.4	283.2
14	200	249.6
15	20.3	220.0
16	218.0	195.2
17	225.6	172.0
18	232.8	153.6
19	240	136.0
20	250.0	121.6

by the ratio of the impedances; that is 600/75 = 8. For example, the T pad 600 ohm R_1 value at 6 dB attenuation can be calculated as 8 × 25 ohm = 200 ohm and R_2 as 8 × 100.5 = 804 ohm. Now for the H pad case each R_1 value will be 100 ohms and each R_2 will be 804 ohms and will provide 6 dB of attenuation. A table of values for R_1 and R_2 versus attenuation for the balanced H pad configuration at 600 ohms is given in Table B.3.

Video noise testing requires a video low-pass filter and a random noise-video-weighting network. The filter and network for EIA RS 250 B are available commercially. The diagram with circuit values is given in Fig. B.5. The components should be mounted on a printed circuit board or a perforated board and placed in an aluminum box for proper shielding. The usual 50 ohm type of BNC connectors should be mounted at each end of the box.

dBmV versus microvolts where 0 dBmV corresponds to 1 millivolt across a 75 ohm impedance is found in Table B.4.

Since the dBm is based on the value of 1 milliwatt of power consumed in a given load, the load has to be specified. A chart of dBm versus voltage for the standard 50 and 75 ohm impedances is given in Table B.5.

A chart showing the equivalent values of voltage standing wave ratio (VSWR), return loss in dB(RL), reflection coefficient, and match ratio is given in Table B.6. The formulas showing the relationships are given in Fig. B.6.

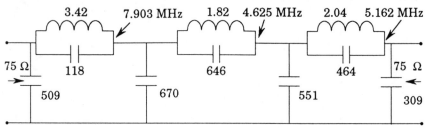

Low-pass filter for video noise measurements $f = 4.2\,\text{MHz}$

L is in microhenrys, C is in picofarads
Q of coils is measured at 5 MHz 80–125 value
(if available a Q meter would be helpful)

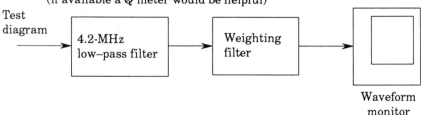

Test diagram

Waveform monitor

Random noise video weighting filter

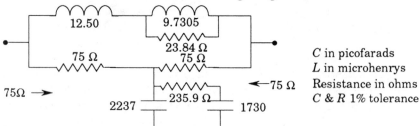

C in picofarads
L in microhenrys
Resistance in ohms
C & R 1% tolerance

f = test frequency
f_1 = 0.270 MHz
f_2 = 1.37 MHz
f_3 = 0.390 MHz

$$\text{Insertion loss (dB)} = 10\log\frac{\left[1 + \left(\frac{f}{f_1}\right)^2\right]\left[1 + \left(\frac{f}{f_2}\right)^2\right]}{\left[1 + \left(\frac{f}{f_3}\right)^2\right]}$$

Figure B.5 Video noise filters

TABLE B.4

dBmV	uV	dBmV	uV
− 60	1.00	+ 5	1778
− 55	1.78	+ 10	3162
− 50	3.16	+ 15	5623
− 45	5.62	+ 20	10,000
− 40	10.00	+ 25	17,780
− 35	17.78	+ 30	31,620
− 30	31.62	+ 35	56,230
− 25	56.23	+ 40	100,000
− 20	100.0	+ 45	177,800
− 15	177.8	+ 50	316,200
− 10	316.2	+ 55	562,300
− 5	562.3	+ 60	1,000,000
0	1000.0		

TABLE B.5

dBm	50 ohm	75 ohm
20	2.24	2.74
15	1.26	1.54
12	0.89	1.09
10	0.71	0.87
6	0.45	0.55
3	0.32	0.39
0	0.22	0.27
− 3	0.16	0.19
− 6	0.11	0.14
− 10	0.07	0.09
− 12	0.06	0.06
− 15	0.04	0.05
− 20	0.02	0.03

TABLE B.6

Reflection coefficient	Match ratio	VSWR	RL(dB)
0.001	1000:1	1.002	60
0.003	316:1	1.006	50
0.010	100:1	1.020	40
0.016	63.1:1	1.032	36
0.032	31.6:1	1.07	30
0.050	19.9:1	1.11	26
0.10	10.0:1	1.22	20
0.16	6.31:1	1.37	16
0.32	3.16.1	1.92	10
0.50	1.99:1	3.01	6
0.79	1.26:1	8.71	2

$$\text{Ref Coef} = \frac{\text{VSWR} - 1}{\text{VSWR} + 1} \qquad \text{RL (dB)} = 20 \, \text{Log} \, \frac{1}{|\text{Ref Coef}|}$$

$$\text{Ref Coef} = \frac{Z_1 - Z_0}{Z_1 + Z_0} \qquad \text{Match Ratio} = \frac{1}{|\text{Ref Coef}|}$$

Z_1 = Load impedance

Z_0 = Characteristic impedance

Figure B.6

Bibliography

Chapter 1

Bartlett, Eugene R., *Cable Television Technology and Operation HDTV and NTSC Systems*, McGraw Hill, New York, 1990.

Byers, T.J., *Electronic Test Equipment Principles and Applications*, McGraw Hill, New York, 1987.

Helfrick, Albert D., and William D. Cooper, *Modern Electronic Instrumentation and Measurement Techniques*, Prentice Hall, Englewood Cliffs, New Jersey, 1990.

Kaufman, Milton, and Arthur H. Seidman, *Handbook for Electronics Engineering Technicians*, Second edition, McGraw Hill, New York, 1984.

Kline, William E., Robert A. Oesterle, and Leroy M. Willson, *Foundations of Advanced Mathematics*, D.C. Heath and Co., Lexington, Massachusetts, 1975.

Schrader, Robert L., *Electronic Communication*, Second edition, McGraw Hill, New York, 1967.

Tektronix Inc. Engineering Staff, "The XYZ's of Using a Scope," Beaverton, Oregon, 1986.

———, "Television Operational Measurements Video and R.F. for NTSC Systems," Tektronix Inc., Beaverton, Oregon, 1984.

Wolf, Stanley, *Guide to Electronic Measurements and Laboratory Practice*, Second edition, Prentice Hall, Englewood Cliffs, New Jersey, 1983.

Chapter 2

Byers, T.J., *Electronic Test Equipment Principles and Applications*, McGraw Hill, New York, 1987

Helfrick, Albert D., and William D. Cooper, *Modern Electronic Instrumentation and Measurement Techniques*, Prentice Hall, Englewood Cliffs, New Jersey, 1990.

Kaufman, Milton, and Arthur H. Seidman, *Handbook for Electronics Engineering Technicians*, Second edition, McGraw Hill, New York, 1984.

Orr, William I., *Radio Handbook*, Howard W. Sams and Co., Indianapolis, Indiana, 1988.

Schrader, Robert L., *Electronic Communication*, Second edition, McGraw Hill, New York, 1967.

Chapter 3

Ballou, Glen, Editor, *Handbook for Sound Engineers, The New Audio Encyclopedia*, Howard W. Sams and Co., Indianapolis, Indiana, 1987.

Benson, K. Blair, and Jerry Whitaker, *Television and Audio Handbook for Technicians and Engineers*, McGraw Hill, New York, 1990.

Buchsbaum, Walter H., Sc.D., *Buchsbaum's Complete Handbook of Practical Electronic Reference Data*, Second edition, Prentice Hall, Englewood Cliffs, New Jersey, 1978.

Jordan, Edward C., Editor in Chief, *Reference Data for Engineers: Radio, Electronics, Computer and Communications,* Seventh edition, Howard W. Sams and Co., Indianapolis, Indiana, 1985.

Tektronix Inc. Engineering Staff, "Spectrum Analyzer Fundamentals," Application Note 26W-7073-1, Tektronix Inc., Beaverton, Oregon, 1989.

Chapter 4

Bartlett, Eugene R., *Cable Television Technology and Operations,* McGraw Hill, New York, 1990.

Benson, K. Blair, Editor in Chief, *Television Engineering Handbook,* McGraw Hill, New York, 1986.

Buchsbaum, Walter H., Sc.D., *Buchsbaum's Complete Handbook of Practical Electronic Reference Data,* Second edition, Prentice Hall, Englewood Cliffs, New Jersey, 1980.

Jordan, Edward C., Editor in Chief, *Reference Data for Engineers: Radio, Electronics, Computers and Communications,* Seventh edition, Howard W. Sams and Co., Indianapolis, Indiana, 1985.

Orr, William I., *Radio Handbook,* Edition 23, Howard W. Sams and Co., Indianapolis, Indiana, 1988.

Schrader, Robert L., *Electronic Communication,* Second edition, McGraw Hill, New York, 1980.

Tektronix Inc. Engineering Staff, "Spectrum Analyzer Fundamentals," Application Note 26W-7073-1, Tektronix Inc., Beaverton, Oregon, 1989.

———, "Television Operational Measurements Video and R.F. for NTSC Systems," Tektronix Inc., Beaverton, Oregon, 1984.

———, "FM Broadcast Measurements Using the Spectrum Analyzer," Application Note 26AX-3582-3, Tektronix Inc., Beaverton, Oregon, 1981.

Terman, Frederick Emmons, Sc.D., and Joseph Mayo Pettit, Ph.D., *Electronic Measurements,* Second edition, McGraw Hill, New York, 1952.

Chapter 5

Bartlett, Eugene R., *Cable Television Technology and Operations,* McGraw Hill, New York, 1990.

Freeman, Roger L., *Telecommunication Transmission Handbook,* Second edition, John Wiley and Sons Inc., New York, 1981.

Jordan, Edward C., Editor in Chief, *Reference Data for Engineers; Radio, Electronics, Computers and Communications,* Seventh edition, Howard W. Sams and Co., Indianapolis, Indiana, 1985.

Kaufman, Milton, and Arthur H. Seidman, *Handbook for Electronics Engineering Technicians,* Second edition, McGraw Hill, New York, 1984.

Rizzi, Peter A., *Microwave Engineering Passive Circuits,* Prentice Hall, Englewood Cliffs, New Jersey, 1988.

Schrader, Robert L., *Electronic Communication,* Second edition, McGraw Hill, New York, 1980.

Tektronix Inc. Engineering Staff, "Spectrum Analyzer Fundamentals," Application Note 26W-7073-1, Tektronix Inc., Beaverton, Oregon, 1989.

———, "Television Operational Measurements Video and R.F. for NTSC Systems," Tektronix Inc., Beaverton, Oregon, 1984.

———, "FM Broadcast Measurements Using the Spectrum Analyzer," Application Note 26AX-3582-3, Tektronix Inc., Beaverton, Oregon, 1981.

Terman, Frederick Emmons, Sc.D., and Joseph Mayo, Pettit, Ph.D., *Electronic Measurements,* Second edition, McGraw Hill, New York, 1952.

Chapter 6

Barker, Forrest, *Communications Electronics,* Prentice Hall, Englewood Cliffs, New Jersey, 1987.

Bartlett, Eugene R., *Cable Television Technology and Operations*, McGraw Hill, New York, 1990.

Freeman, Roger L., *Telecommunication Transmission Handbook*, Second edition, John Wiley and Sons, Inc., New York, 1981.

Jordan, Edward C., Editor in Chief, *Reference Data for Engineers: Radio, Electronics, Computers and Communications*, Seventh edition, Howard W. Sams and Co., Indianapolis, Indiana, 1985.

Kaufman, Milton, and Arthur H. Seidman, *Handbook for Electronics Engineering Technicians*, Second edition, McGraw Hill, New York, 1984.

Lindberg, Bertil C., *Troubleshooting Communications Facilities*, John Wiley and Sons, Inc., New York, 1990.

Chapter 7

Barker, Forrest, *Communications Electronic,* Prentice Hall, Englewood Cliffs, New Jersey, 1987.

Bartee, Thomas C., Editor in Chief, *Digital Communications,* Howard W. Sams and Co., Indianapolis, Indiana, 1986.

Byers, T.J., *Electronic Test Equipment Principles and Applications,* McGraw Hill, New York, 1987.

Helfrick, Albert D., and William D. Cooper, *Modern Electronic Instrumentation and Measurement Techniques,* Prentice Hall, Englewood Cliffs, New Jersey, 1990.

Kaufman, Milton, Arthur H. Seidman, *Handbook for Electronics Engineering Technicians,* Second edition, McGraw Hill, New York, 1984.

Lindberg, Bertil C., *Troubleshooting Communications Facilities,* John Wiley and Sons, Inc., New York, 1990.

Martin, James, *Local Area Networks,* Prentice Hall, Englewood Cliffs, New Jersey, 1989.

Peebles, Peyton Z. Jr., Ph.D., *Digital Communication Systems,* Prentice Hall Inc., Englewood Cliffs, New Jersey, 1987.

Wist, Abund Ottokar, Z.H. Meiksin, *Electronic Design of Microprocessor-Based Instruments and Control Systems,* Prentice Hall, Englewood Cliffs, New Jersey, 1986.

ABOUT THE AUTHOR

Eugene R. Bartlett is a cable industry consultant.
Formerly, he was a senior staff engineer with Tele-Media
Corporation, a cable company in Pleasant Gap,
Pennsylvania. He holds a MSEE from Northeastern
University and is the author of *Cable Television
Technology and Operations* (1990).

Index

Accuracy, 5
of measurement, 6
ADC (*see* Analog-to-digital converter)
AFC (*see* Automatic frequency control, radar systems)
Alternating current (ac), 51
measurement, ammeter, 51
measurement power, 52
meters, 22
Alternating voltages radio frequency, 99–101
American standard code for information interchange (ASCII), 176, 177, 215, 217, 223
Ammeter, 49–52
alternating current ammeter, 49, 51, 52
clamp on ammeter, 49–52
direct current ammeter, 49, 50
Amplifier cascade, 181
Amplifiers, audio, 74–76
distortion, 76
gain, 74, 76
power, 74–76
Analog instrument, 6
electromechanical, 6
readings, 6
scales, 6
Analog-to-digital converter (ADC), 219–221
Antennas-circuits, 100
Arithmetic mean, 10
ASCII (*see* American standard code for information interchange)
Audio signals, 71–90
crosstalk, 85
gain, 74, 75
hum, 82, 83
noise, 82–85
pink noise, 85
power, measurement, 74, 75
stereo, separation, 85–90
volume, 73
waveform, 72

Automatic frequency control (AFC), radar systems, 164
Average (mean) deviation, 10

BER (*see* Bit error rate)
Bessel null method microwave, 156, 157
modulation test, 135–138
Binary numbers, 213–218
binary zero and one, 214
Bit error rate (BER), 160, 161, 230, 232
microwave systems, 160, 161
tester, 230, 232
Block converters microwave, 152
Break-out-boxes, 226–228
BTSC (*see* Television Stereo Broadcast Channel)
Byte, 213, 217, 220
8-bit byte, 217, 224

Cable system leakage, 204–208
FCC rules, 204–206
frequency offsets, 205
leakage limits CLI, 206–208
test methods, 204–208
Cable systems, types, 171
telephone, 171, 172
television, 171
Capacitance, capacitors, 53–57
high frequency capacitance, 58
measurement of:
bridges, 58–60
meter, 57
types of, 60
Capacitance cable, 191
Capacitance shunting, 178
Carrier-to-noise microwave systems, 152–154
Cathode-ray tube (CRT), 27–28
CATV (*see* Community antenna television)
CCIR (*see* International radio consultative committee)
CLI (*see* Cumulative leak index)

CMOS (*see* Complimentary metal oxide semiconductor)
Coaxial cable testing, 190
   continuity test, 190
   fault test, 191, 192
   frequency response, 192–197
Coaxial cables, 178
   attenuation, 180
   capacitance, 179
   constants, 179
   resistance, 179
Community antenna television (CATV), 180
Compact disk players, 96, 97
Complimentary metal oxide semiconductor (CMOS), 213
Composite triple-beat, 198–202
Crosstalk, audio systems, 85, 87, 88
CRT (*see* Cathode-ray tube)
Cumulative leak index (CLI), 205
   calculation, 206, 207
   FCC rules, 205, 206
   measurements, 206–208
   records, 205
Current flow, 49
   high current, 50

DAC (*see* Digital-to-analog converter)
DC (*see* Direct current)
Decibels (dB), 13–16
   bel unit of, 13
   electrical parameters, 2
   1 millivolt reference dBmV, 14
Deutsche Industrie Norm (DIN), 78
   audio noise, 83
   audio tape recording, 92
Deviation from mean, 10
   average (mean) deviation, 10
   standard deviation, 11, 12
DFB (*see* Distributed feedback laser)
Digital bus lines, 223
   EIA RS 449, 223, 231
   GPIB, 225–229
   IEEE 448, 225, 226, 229
   RS-232, 223, 224, 226, 228, 229
Digital codes, 219
   ASCII code, 176, 177, 217
Digital communication systems:
   telegraph, 175
   telephone T-1, 230
Digital meters, 7
   multimeter, 24, 25–27, 48–52
   voltmeter, 24, 27, 45–47

Digital multimeters, 48
Digital signal testing, 150, 161
Digital-to-analog converter (DAC), 220
DIN (*see* Deutsche Industrie Norm)
Direct current (dc):
   ammeter, 49, 51, 52
   meters, 22
   voltage, 46
Directional coupler, 103, 104
Distribution curve, 8–12
   normal (Gaussian distribution), 8
Distributed feedback laser (DFB), 188
Doppler effect radar systems, 162
Dual-tone multifrequency (DTMF), 174
   telephone dialing, 174

ECl (*see* Emitter-coupled logic)
8-bit byte, 217, 224
   (*See also* American standard code for information interchange)
Electrical parameters:
   current, 1
   decibel, 2
   frequency, 2
   period, 2
   power, 1
   resistance, 1
   time varying, 2
   voltage, 1, 45, 46
      alternating, 47, 48
      direct current, voltage, 46
      high voltage, 46
Electromechanical meter, 43
Emitter-coupled logic (ECL), 213
Energy storage elements, 4
   capacitance, 4
   inductance, 4
EPROM (*see* Erasable programmable read only memory)
Equalized conditioned lines, 172
Erasable programmable read only memory (EPROM), 222
Error in measurements, 17
   accuracy, 5–6
   tolerance, 17
Error probability, 11
Eye pattern, 231

Fabry Perot laser, 188
Federal Communications Commission (FCC):
   broadcast rules, 108–112
   cumulative leak index rules, 204–206

Federal Communications Commission
(FCC) (*Cont.*):
    frequency standards, 107
Fiber-optic cable communications,
    186–188, 209
    cable parameters, 209
    construction, 187
    systems, 189
Fiber-optic transmitters, in lasers, 188
Flip-flops, 213
    D latch, 214
    J-K, 213, 214
    reset-sat RS, 213
Frequency, 2
    hertz (Hz), 3
    resonant frequency, 5
Frequency modulation stereo service, 132
    testing, 133–138
        modulation testing, 135
Frequency response cable systems,
    192–194
Frequency shift keying (FSK), 177, 215
    cable LAN modulation, 183

Gallium arsenide field effect transistor
    (GaAsFET), 143
Gauss, Karl Freidrich, 8
Gaussian curve, 8, 9, 10
Grounding and bonding, audio systems,
    88, 89

Hay bridge, 61
Heading flasher, 164
Hertz, Rudolph Heinrich, 3
Hertz (Hz), measurement, 3
Hum in cable systems, 203
Hz (*see* Hertz, measurement)

IC (*see* Integrated circuits)
Impedance:
    circuit loading, 5, 63, 64
    high impedance, 5
    input impedance, 63, 65, 66
    mismatch impedance, 67–69
    output impedance, 63, 65, 66
    resistive impedance, 5
    terminal impedance, 5
Impedance audio line, 72
    impedance balance line, 72
Impedance microwave measurement of,
    144
Inductance, 53–62
    bridge, 60–62

Inductance (*Cont.*):
    cable, 191
    loading of lines, 178
    meter, 60
Inductance loading of lines, 178
Insertion loss CATV system, 182
Instruments, basic electrical, 18
Integrated circuits (IC), 214
    large-scale integrated circuits (LSI),
        214
    very large-scale integrated circuits
        (VLSI), 214
Intermodulation distortion:
    audio, 81
    composite triple-beat, 198, 199
    third-order distortion, 198–202
International radio consultative commit-
    tee (CCIR):
    distortion standards, 78, 81
    noise, 83
    tape recording, 91, 92
    phonograph, 95
Interstage trim networks CATV, 195

Kelvin bridge, 173
Klystron, 141

LAN (*see* Local-area-network)
Large-scale integrated circuits (LSI), 214
Lasers:
    compact disk players, 96
    fiber-optic transmitters, 188
Light-emitting diode, 221
Liquid crystal display (LCD), 42, 221
    graphs, 42, 43
    screens, 42, 43
Line of site microwave link, 147
    transmit/receive measurements,
        148–150
Local-area-network (LAN), 180
    carrier-to-noise test (C/N), 186
    description, 184–185
    ethernet, 185
    map/top, 185
    testing, 231, 232
    token ring, 185
Logic analyzer, 228, 229
Logic circuits, systems:
    complimentary metal oxide semicon-
        ductor (CMOS), 213
    emitter-coupled logic (ECL), 213
    flip-flops, 214, 218
    logic functions, 214–218

Logic circuits, systems (*Cont.*):
  negative true, 214
  positive true, 214
  transistor logic (TTL), 213
Logic probe, pulser, 226–228
LSI (*see* Large-scale integrated circuits)

Magnetron, 141
Maxwell bridge, 61
Meters:
  accuracy, 25–27
  alternating current (ac), 22
  circuit loading, 19–21
  computer controlled, 25
  direct current (dc), 22
  electromechanical, 18
  electronic, 24–27
  full scale, 19
  high impedance, 24
  multimeter, 24
  ohmeter, 23
  ranges, 19
  resistance, 19
  sensitivity, 18
Microphones, carbon, 173
Microwaves, 103
  frequency measurement, 145
  modulation, 146, 147
  network analyzer, 146
  polarization, 152
  power meter, 143
  spectrum analyzer, 145
  systems, 139–143
    communications, 141
    generators, 141–143
    testing, 152–154, 157–160
Modulation:
  amplitude (AM), 109, 110
  measurement, 109, 110
  frequency (FM), 109
  television (NTSC), 111–133
Multimeter, 24, 25–27, 48–52

National Association of Broadcasters
    (NAB), 92, 95
  audio tape recording, 92
  phonograph, 95
National Bureau of Standards (NBS), 16,
    101
  calibration standards laboratory, 16,
    17
  frequency standards, 107, 108
National television study committee, 49

NBS (*see* National Bureau of Standards)
Network analyzer, 42
  microwave, 146
Network topology, 189
Noise:
  carrier-to-noise ratio (C/N) cable
    systems, 202–204
  microwave systems, 152–154

Ohmeter, 23
Optical power meter, 209
  measurements, 210
  path loss, 210
Optical time domain reflectometer
    (OTDR), 211
  cable-fault testing, 211, 212
Oscilloscopes:
  accuracy of measurement, 34–36
  cathode-ray tube, 27, 28, 36
  development of, 28, 36
  dual beam-trace, 31, 32
  measurements, 29–31
  storage, 33, 34
  triggering, 33
OTDR (*see* Optical time domain
    reflectometer)

Parabolic antennas, 162
  radar systems, 162
  satellite systems, 150
Parity word, 217
PCM (*see* Pulse-code modulation)
Period, 2–3
Phonograph recording, 90, 94–96
Pilot carrier, FM stereo, 132, 133
Pink noise, audio signals, 85
Plan position indicator (PPI), 165
Polarization:
  microwaves, 152
  satellite systems, 155, 156
Power factor, 53
Power meter, RF power, 101–103
PPI (*see* Plan position indicator)
Preemphasis, audio tape recording, 92
Probes measuring:
  low-capacity oscilloscope, 63–65
  word recognizer, 234
Programmable only memory (PROM), 222
PRR (*see* Pulse repetition rate)
Pulse-code modulation (PCM), 215, 217
Pulse repetition rate (PRR), 163, 164

Q-meter, 62, 69

Radar, 140, 161, 162
  maintenance, 166–170
    diagnostics, 169, 170
    precautions, 166
    testing procedures, 167–169
  marine navigational, 163
Random access memory (RAM), 219, 232
Resistance, 53
  high resistance, 56, 57
  low resistance, 55, 56
  measurement, 54–57
  ohmmeter, 54, 55
Resonance:
  condition, 5
  frequency, 5
Reverse amplifiers, 183
Rotary dialer, 174

Satellite communication, 140, 150
  measurements, 151–161
  uplink downlink, 150, 151
Second-order distortion, 199, 202
SI (see System of international units)
Signal distortion cable/LAN, 197
Signal-level meter, 194
  cable system, 195–197
Signal-to-noise ratio, 13
Significant figures, 7
  digits, 7
  zeros, 7, 8
Slotted transmission line testing, 105–107
Society of Motion Picture and Television Engineers (SMPTE), 78
Spectrum analyzer:
  measurements, 40, 41
  operation, 39, 40
  principles, 38–40
Standard deviation, 11, 12
Standing wave ratio (SWR) microwave, 144
  voltage standing wave ratio (VSWR), 144, 145
System of international units (SI), 15, 16

Tape recording, audio tape recording, 91–94
Taps, CATV, 181
Telephone systems, 171, 172
Teletypewriting (TTY) code, 176
  ASCII, 176, 177
Television broadcast transmitter, 111

Television carrier frequency tests, 130–132
Television signal testing:
  chrominance-to-luminance delay inequality, 122, 123
  chrominance-to-luminance gain inequality, 119–123
  differential gain, 123, 124
  differential phase, 124, 125
  field time distortion, 119
  gain-frequency response, 122
  insertion gain, 117
  line time distortion, 118
  noise, 125, 126
  short time distortion, 118
Television Stereo Broadcast Channel (BTSC), 128–130
Television video test signals, 113–115
Television waveform monitor, 36–38
  video waveform, 36
  waveform monitor graticule, 36
THD (see Total harmonic distortion)
Thermocouple instruments, 101
Third-order distortion, 198–202
Time domain reflectometer, 41–42
  measurements, 41
  operation, 41
Total harmonic distortion (THD), 77–81
  distortion analyzer, 77–81
Transfer function, audio amplifier, 76
Transistor logic (TTL), 213
Transmission line RF 100, 103–107
  slotted line measurements, 105
Transmit/receive radar, 164
  anti-transmit/receive switch, 164
  transmit/receive switch, 164
Trapping filters CATV system, 182
Travelling wave tube (TWT), 142
Truth tables, 214–217
TTL (see Transistor logic)
TTY (see Teletypewriting code)
Two-way cable systems, 183
  system alignment, 195–197
TWT (see Travelling wave tube)

Universal asynchronous receiver transmitters (UART), 224

Vector scope, 37, 39, 113
  graticule, 116
Vertical interval test signals (VITS), 113
Very high frequency (VHF), 48

Very large-scale integrated circuits
    (VLSI), 214
VHF (*see* Very high frequency)
Video signals, 111–133
    composite video, 112, 113
    test signals, 37
Video waveform, 36
VITS (*see* Vertical interval test signals)
VLSI (*see* Very large-scale integrated
    circuits)
Voice frequencies, 73
Volume unit (VU), 73
Voltage standing wave ratio (VSWR),
    105–107

Voltmeter, 24, 27, 45–47
VSWR (*see* Voltage standing wave ratio)
VU (*see* Volume unit)

WAN (*see* Wide-area-network)
Wattmeter, 52, 53
Waveform monitor, 113
    graticule, 115
    termination resistor, 116
Waves-waveform:
    sine, 3
    television, 36
Wheatstone bridge 55, 56, 173
Wide-area-network (WAN), 231, 232